T0239183

The Science of Learning
Mathematical Proofs

An Introductory Course

The Science of Learning Mathematical Proofs

An Introductory Course

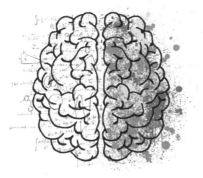

Elana Reiser
St. Joseph's College, USA

World Scientific

NEW JERSEY · LONDON · SINGAPORE · BEIJING · SHANGHAI · HONG KONG · TAIPEI · CHENNAI · TOKYO

Published by

World Scientific Publishing Co. Pte. Ltd.

5 Toh Tuck Link, Singapore 596224

USA office: 27 Warren Street, Suite 401-402, Hackensack, NJ 07601

UK office: 57 Shelton Street, Covent Garden, London WC2H 9HE

Library of Congress Cataloging-in-Publication Data

Names: Reiser, Elana, 1979–   author.

Title: The science of learning mathematical proofs : an introductory course /
   Elana Reiser, St. Joseph's College, USA.

Description: New Jersey : World Scientific, [2021] | Includes bibliographical references and index.

Identifiers: LCCN 2020044703 | ISBN 9789811227677 (hardcover) |
   ISBN 9789811225512 (paperback) | ISBN 9789811225529 (ebook for institutions) |
   ISBN 9789811225536 (ebook for individuals)

Subjects: LCSH: Proof theory.

Classification: LCC QA9.54 .R45 2021 | DDC 511.3/6--dc23

LC record available at https://lccn.loc.gov/2020044703

British Library Cataloguing-in-Publication Data

A catalogue record for this book is available from the British Library.

For any available supplementary material, please visit
https://www.worldscientific.com/worldscibooks/10.1142/11973#t=suppl

Desk Editor: Liu Yumeng

Printed in Singapore

This book is dedicated to all of my students, past, present, and future.

Preface to Students

This introduction to mathematical proofs course is a bridge between the math that you are used to — mainly calculations — and more abstract mathematics. This may completely change the way that you view mathematics and it will take some getting used to. To help prepare you for this transition, you will first learn some background information on topics such as changing your mindset, learning the best ways to learn, and learning how to work well with a group. We will break down many false misconceptions about math, such as making a mistake is something to be ashamed of or being smart means being fast. These habits of mind will leave you in a prime position to get the most out of the proof chapters, once we get to them later on. The best way to introduce this course is from a former student who has taken it. Below is the student's impression of the course.

> This class is different than any other math course that I have taken. At the beginning of the course we go over how important having a growth mindset is. It is easy to get discouraged in this class because it's different than what you're used to. Take those lessons seriously and believe it; it's normal to make mistakes; in fact, mistakes are essential to learning. Turn those mistakes into an opportunity for growth, rather than a discouragement. As you begin to write proofs, class participation and working with a group is important. Hearing all of the different perspectives and opinions is very helpful in this class. The material learned in the course can be applied to almost everything. The importance of problem solving and critical thinking are valuable skills to take with you beyond a school setting.
>
> Elana Reiser

Preface to Professors

The idea for this book was a slow process. I have taught the introduction to proofs course for many years and have always struggled with switching students from thinking of mathematics as a procedure and into a creative process. So many times in this class and in upper level math courses I have heard my students say "I am not good at proofs" or "I hate proofs." And each time it broke my heart a little.

I have always liked using group work during class so that students can bounce ideas off of each other and work together to find solutions. I am a strong believer in the child psychologist Lev Vygotsky's learning theory that "what a child can do in cooperation today he can do alone tomorrow." But I found that even working in groups, students would make one attempt at a proof, not get it, and then give up. A major problem was a lack of willingness to struggle with a problem. If they didn't see the answer right away, they gave up. I attempted many approaches to try to help my students see that they needed to struggle through a problem. One summer I enrolled in Stanford Professor Jo Boaler's online workshop entitled *How to Learn Math for Teachers* and it clicked. My students had fixed mindset thinking whereas I wanted them to have a growth mindset. Boaler introduced strategies to make this change and I have since read many books on this topic, attended many workshops, and participated in many discussions, both online and in person. There is currently a strong movement to make these kinds of changes in the K-12 education world, but not as much at the post-secondary level. This introduction to proofs course is exactly the point where students need an intervention, before they get to upper level math courses that are extremely proof heavy. I strongly believe that introducing these concepts to students before they start working on mathematical proofs will help them persevere and not give up as quickly if they don't understand how to write a proof right away. Here are some quotes from former students who have taken this course.

"Going over those topics at the beginning of this semester gave me the motivation to be able to do problems I wouldn't know how to do."

"It changed my approach when it comes to solving difficult problems."

"It made me look at math in a different way than just calculating numbers and doing questions with the answer being a number."

"This proof course helped me [have] the prior knowledge before going into higher level math courses. It made the higher level courses easier to comprehend."

There are some chapters in the beginning of this book that don't even have proofs in them. To start off, your students will learn the difference between a fixed mindset and a growth mindset. This helps prime them for proof writing. Carol Dweck is a psychologist who first coined the term 'growth mindset'. Here are a few reasons from Dweck's book *Mindsets* why fixed mindset thinking can be harmful:

- People who have a fixed mindset think that they can't make a mistake because if they do then that is a sign of weakness or that they are not smart.

- People who have a fixed mindset think that if you have to put effort into something and struggle with it then that means you are not smart; things just come easily to people who are smart.

- It is harmful for someone with a fixed mindset to be labelled as smart because then they have to keep re-proving that they are smart. If they encounter a failure then that must mean they are not smart anymore.

- As a result, people with a fixed mindset may put less effort into studying for a test or performing an assignment because that means that if they fail they can always blame their lack of effort rather than thinking it is because they are not smart.

For these reasons, and so many more given in more detail in Dweck's book, it is important for students to at least be made aware of how harmful having a fixed mindset can be. Once they have this awareness they can begin to work towards developing a growth mindset. I have provided many exercises throughout this book to help this development.

Different students take this course at different points in their studies but the ideal time is generally as freshmen or sophomores right after taking Calculus 1.

Elana Reiser

Contents

Chapter 0

Pedagogical Notes for Professors

0.1 Introduction

In this chapter I will provide suggestions on how to implement many of the ideas given in this book. These will be research-backed suggestions that I use in my classroom, but feel free to pick and choose what works best for you. Everybody has their own teaching style, so a technique that I like may not work for you.

In this course the first day of class is very important because it will set the tone for the rest of the semester. Teaching students about growth mindset will be pointless unless you also believe in it and set up your classroom to promote growth mindset thinking. To that end, there are some teaching practices that I use to help get across the idea that I care about my students and want them to succeed. On the first day of class I have students fill out a "getting to know you" questionnaire. Some of the questions I ask on there is what their favorite color is and what their favorite band is. I print the first quiz on paper that corresponds to their favorite color. This is a simple but caring gesture. When we get to classes where students spend time working in groups I sometimes play soft background music. I use some of the bands that they have listed as their favorites to play in class. As part of their first homework assignment I ask students to visit me during office hours. I find this useful for several reasons and I share these reasons with my students. First, it shows them where my office is and starts to make them feel comfortable going there. They know that they can always come to office hours if they need extra help. Second, it also gives me a chance to briefly chat with each student individually to start to get to know them better.

0.2 Homework

The biggest, most useful, but also very time consuming growth mindset practice I use is to allow multiple submissions on homework. This shows students that mistakes are learning opportunities and that they don't have to do a problem perfectly on the first try. The way that this works is by having students submit their homework digitally. I have tried having students type up their solutions in the past but they have found this very time consuming and I actually prefer to see work done in their own handwriting. Instead, I ask students to scan their work. There are digital scanner apps that make the scanning process simple and for students without a smartphone or tablet, we have a scanner at the library that sends a digital file to the student's email address. I do make clear my expectations on the first day of class by showing students an example of a page that has been digitally scanned and a page that a student just used the camera on their phone to take a picture of. The difference in how much neater and easier to read the scanned page is makes it clear why I am asking them to take this extra step. For the most part they do use the digital scanner throughout the semester.

Every homework has a due date and students are allowed to hand in homework any time before the due date. Once they submit I will read it over and give feedback on what they got correct and let them know if they got something wrong. If a problem is incorrect or if they are stuck I will give them a hint or ask a guiding question to help them get to the next step. They are allowed to submit as many times as they like up until the due date. I no longer accept submissions once the assignment is due. There are some students who go all in and take advantage of this help. Others may submit one extra time and some only submit once on the due date. As I said, this can be time consuming but I truly think it is worth it. You start to see the same types of mistakes so it doesn't take much time to give the feedback. I just spread the grading out by checking in many times each day. Once I have started grading homework this way I can't imagine ever going back to just collecting it once. This practice also helps eliminate cheating because students don't need to have the problem worked out perfectly the first time.

0.3 Chapters in book

In Chapter 1 students are introduced to growth mindset and how developing one can help them persevere through writing mathematical proofs. Some additional resources you may be interested in using during class are video clips. Khan Academy has an online program called LearnStorm that introduces students to growth mindset. You don't have to use their full program, but there are some videos within it that may be helpful. I also like to have students watch a video on famous failures. There are many options available on YouTube using that search term.

Chapter 2 is on cooperative learning and helping students work together in groups. There are several tasks presented in this chapter designed to help students practice some of the skills needed for a successful group. You may decide to spread these activities out over several days rather than doing them all on the same day. I generally mix the Chapters 1 and 2 material and spread it out over several class periods.

In Chapter 3 students set a SMART goal. It can be helpful to have them check in at various points throughout the semester to see how they are progressing with this goal.

Chapter 4 covers various logic topics. I have this spread out over three class periods mixed with Chapters 3 and 6.

In Chapter 5 we cover problem solving. I also spread this topic over three class periods. The first class we do the exercises given in the chapter. One additional homework I usually give is for students to pick one of the problems we did in class and give it to a friend or family member. If the friend or family member gives up or gets frustrated, students are asked to share some of the ideas we have learned about a growth mindset and how making mistakes is okay. The second class we have a game day. Thinkfun makes a lot of great logic and problem solving games. I use the game they make called Chocolate Fix. You are given a set of clues and have to figure out how the chocolate pieces fit onto the tray. Similar logic games but with different contexts include Gravity Maze and Rush Hour. Qwixx is another good game. There is a set of six dice that you take turns rolling. Each player can sum any two dice and cross off that sum on their scorecard. The scorecard has four rows of numbers. Once you cross off a number you can't cross off a number listed in that row earlier. This way it becomes a game of strategy. Other more classic games include Set and Mastermind. Ghost Blitz is a game similar to Set with various items set in the middle. A card is turned over with multiple items shown in various colors and players have to be the first to grab either the item shown on the card that is the same color as on the card or an item not on the card that is a different color from all other items on the card. On the third day we do an escape room that I created. See Appendix C for more details on this.

In Chapter 6 we go over various study techniques that help students understand the best ways to learn. As stated in this chapter, all of these ideas came from a combination of the following books: *Make it Stick* by Brown, Roediger, and McDaniel; *A Mind for Numbers* by Barbara Oakley; and *How We Learn* by Benedict Carey. I would highly recommend reading one or more of these books. One of the ideas presented in this chapter is for students to take breaks. I incorporate short breaks into some classes. A few ideas you may be interested in trying are mindfulness or meditation type exercises. I got a set of yoga cards that includes breathing exercises and let students pick one to do. One example of a breathing exercise is to stretch your fingers out on one hand and use your pointer finger from the other hand to trace your spread out fingers. As you go up breathe in and as you go down breathe out. This is a simple, quick exercise that can be used as a short break to refocus the class.

The goal of Chapter 7 is to prepare students mentally for starting proofs. I included this chapter because it is very important to combat any prior experiences students have had with proofs that may have given them a false, bad sense of how writing proofs feels like or looks like.

The remaining chapters are when students learn how to write proofs. Chapters 8 through 18 focus on number theory as the topic for teaching mathematical proofs. Chapter 19 covers Calculus proofs. There are not any new proof techniques in this chapter, but it utilizes many of the previously learned techniques and frames them in a new context. One video that is great to show for this class is by mathematician Grant Sanderson. He has a collection of videos that give a visual representation of many different math topics. The one I use is on delta-epsilon proofs [Sanderson, 2017]. Chapter 20 is a mixed review with proof questions from Chapters 8 through 19. Possible solutions are also included. Next, I will discuss some current research on how the classroom setup can be conducive to aiding proof writing, as well as some possible structures to use during class time and how assessment fits in.

0.4 Thinking classrooms

A Canadian math professor named Peter Liljedahl [2016] developed the idea of building thinking classrooms. He has a list of fourteen practices that can be used to help achieve this. The top three suggestions he gives to get started are to give students highly engaging tasks, to use visibly random grouping, and to have students working on vertical non-permanent surfaces. Liljedahl's research has found that these practices help to get students thinking and keep students engaged. Highly engaging tasks are important because not all tasks are conducive to group work. A group-worthy task should be productively uncertain, where students do not necessarily know how to get to the "answer" the teacher is looking for, and require complex problem solving. This allows students to debate and work through issues together while learning from any mistakes they make [Cohen and Lotan, 2015]. Proofs are great examples of highly engaging tasks.

Random grouping is important because it generally leads to hetero-geneous groups. This allows for students of different backgrounds to work together so that multiple viewpoints can be heard. The visible component is also important because this shows students that the teacher didn't make groups for a specific purpose, such as that each contains a high achiever and a low achiever. Benefits to visibly random grouping include: students becoming more agreeable to work in the group they are placed in, elimination of social barriers, less reliance on teacher and more reliance on other students, and more engagement and enthusiasm in class [Liljedahl, 2016]. There are various ways to achieve visible random grouping. If you use a learning management system such as Canvas it should be able to randomly form groups and you can display the random grouping on a projector as it happens. I have used this method in class and the students really enjoy it because it gives them a sense of excite-ment. Some classes I gave them the option of forming their own groups or using the random grouper and they always chose the random grouper. Another option is to use a website such as this one to create random groupings: https://www.randomlists.com/team-generator.

Vertical non-permanent surfaces refer to whiteboards, blackboards, windows, or any surface you can write on and then erase. They are bene-ficial for a variety of reasons. In one study students from five different classes were assigned a work surface for problem solving of either a wall mounted whiteboard, a whiteboard to use at their desks, a large paper taped to the wall, a large paper to use at their desks, or their own notebook to use at their desks [Liljedahl, 2016]. It was found that students using whiteboards (non-permanent surfaces) were more eager to start problem solving, had more discussion, had more participation, were more persistent, and included more non-linear thinking than the students using permanent surfaces. The non-permanence gives students permission to make mistakes because they can easily be erased. The study actually found that erasing mistakes on the whiteboards rarely occurred and students instead moved to another part of the board to start a new idea. This led to the higher non-linearity of their problem solving when using whiteboards. When using a permanent surface students were reluctant to write anything down unless they knew for sure that it was leading to a solution. The fact that all students are using a single workspace rather than having each student individually writing in their notebook promotes collaboration.

When the non-permanent surfaces are vertical and placed around the classroom students are standing up as they work. If you need convincing of why this is helpful then do a quick experiment in one of your classes where students work in groups standing up versus sitting down. It is amazing how much more students are engaged when they are standing up. Every student is participating. The reason for this is that standing discourages anonymity. Students standing up to work also makes them more likely to seek help from other groups before going to the teacher [Liljedahl, 2016]. The work of all groups is visible to everyone so it is easy to see what other groups are doing. Another tip that Liljedahl gives to go along with this is to only give one marker per group. This prevents students from going off on their own and forces the group to work together.

0.5 Classroom structures

Next, I will discuss some ideas for use during class time. One cooperative learning structure that I frequently use is called Think, Pair, Share. In this structure students are given some time to think about or solve a problem individually. Then they meet with a partner to discuss their answers. Finally, the class meets together to discuss what the pairs found. I use this structure any time I have students write an answer to a reflection question in class. I also use a timer for the think and pair parts, generally giving two minutes for each portion. This keeps the class on task. Allowing time for students to write their answers independently is important to allow them to think and process how to answer a question before forcing them to articulate it. It also ensures that each student is putting their own thought and effort into the answer. The discussion between pairs allows all students to be heard. We don't have time for everyone to share to the whole class, but at least every student is sharing their thoughts with somebody else. It is also a way to ensure full class participation in a timely manner.

Another cooperative learning structure I sometimes use is called Four Corners. In this structure you would post a problem in each corner of the room and have students working in groups to solve a problem. They would then rotate and move to the next problem and continue like this until all students have looked at all problems. This structure is another way to get students out of their seats. As we discussed earlier, this is beneficial to making sure all students are participating and engaged. This structure is especially useful for exercises such as 8.11 that have four possible proofs and students need to decide if each one is correct or incorrect.

0.6 Assessment

Finally, I will share some ideas for assessment. This is another topic that is getting a lot of discussion in the K-12 math world, but not as much in higher education. In a college level course it is standard practice to give tests and a final and count that as the majority of the grade. Along with the increasing awareness of growth mindset, there is also tied to it the idea of standards-based grading as an alternative assessment method. They fit well together because standards-based grading gives a snapshot of where each student is on their learning journey with the idea that they can always improve. Think about what you believe the purpose of assessment is and see if the types of assessments you use match with that definition. The way that I view it is that the course goals I list on my syllabus are what a student at the end of the semester should come out of my class knowing. If I give a test towards the beginning of the semester and a student does poorly on it, but then by the end of the semester, through hard work and practice, that student is able to understand the concepts that he or she initially did poorly on, is it fair to punish that student by including the bad grade as part of the final grade? I would love to have a fully standards-based grading system in place, but in reality I have a mixture of that along with traditional grading. Earlier I discussed how homework is graded. Allowing students to hand it in multiple times with feedback given between each submission is a way to allow students to show growth, but I do have a deadline because otherwise I wouldn't be able to go over the problems together in class. The way that I grade each proof is on a scale from 0 to 3. A score of 0 is given if the student did not attempt the problem. A score of 1 is given if the student has some good work in there but didn't really get the concept. A score of 2 is given if the student is getting there but doesn't quite understand the proof. A score of 3 is given if the student got the proof correct, with possibly a small calculation error. I do sometimes also give half points.

Once we get to Chapter 8 when the proofs start I give short, frequent quizzes. Generally they will be one fairly basic proof. I also have students give a rating on how well they think they understand whatever topic the quiz is on. Just a quick question like: I feel that I understand (whatever the topic is) completely, somewhat, or not yet. When I grade these quizzes I do not put the numeric grade on them, only feedback. I do put the numeric grade in the online gradebook that they have access to. The reason for not including it on the quiz is that students tend to focus on only the number and not the comments. By leaving the grade off, this gives students the chance to read the comments and reflect on their work. Additionally, after quizzes and tests students reflect on what they got wrong. Here is a sample reflection sheet to give students to fill out.

Reflection: Why I missed the original
Didn't understand
Thought it was right
Skipped a step
Not sure
Guessed
Careless mistake _____
Other: _____

What I can do in the future to prevent this from happening again:

At the end of each proofs chapter I put a reflection question that can be used as an exit ticket or a chance for a reflection on what they learned that day. There are two types of reflection questions. One asks students to think about any mistakes they made or heard another student make and how correcting that mistake helped them learn. The other type of reflection is to summarize what they learned that day and what questions or concerns they have about the material.

Another assessment tool that I use is portfolios. I use Google Sites to create a template for the portfolio and students use that as a starting point in creating their own. For each proof type I create a page that has a learning standard. For example, for proof by contraposition the standard states: "I understand and can use the proof technique of contraposition." Students have to upload evidence for each standard to show that they are eventually proficient at it. I require them to upload one piece of evidence around the time we first learn the topic. This first post does not need to show proficiency, but then they have to upload a second piece of evidence later on in the semester at the point when they are proficient. They can post more than twice. Students get many chances to practice each topic since the homework is spiraled. I also provide a sample page showing what the entries could look like. These are the directions included on the front page.

Each standard has its own page. For each standard provide evidence to show you met it. Examples of evidence can include homework solutions, questions from a test, a video explaining a problem, a visual explaining the concept, a written explanation of concept, or any other way you can think of to show that you met that standard. Each standard should follow the following format.

Around a week after learning topic:

- Rating on how well you understood the topic (0 = no understanding, 1 = a little understanding, 2 = partially proficient, 3 = just about proficient, 4 = proficiency)
- Evidence of understanding up to this point. You do not need to show proficiency at this point, just show what understanding you have

At some later point in the semester:

- Rating on how well you understood the topic
- Evidence to show proficiency (your goal is to get a 3 or 4 proficiency rating)

- Describe your growth on this standard
- Describe this concept to a 15 year old

You can give evidence multiple times to show growth but a minimum of two are required, once right after you learn the topic and once when you feel you have reached proficiency.

Chapter 1

Brain Growth

1.1 Mindsets

Before diving into this chapter, let's start by taking a quiz. Answer each question as true or false.

(1) I think that people are born with a certain level of intelligence and no matter what they do that intelligence level can't be changed.

(2) I think that a person is either born a math person or not.

(3) I think that smart people just get it and they don't need to put much effort into studying.

If you answered "true" to most of these questions then you likely have what is called a fixed mindset. If you answered mostly "false" then you have what is called a growth mindset. People with a **fixed mindset** think people are either born smart or not smart and any type of struggle or failure is seen as a sign of weakness, a sign that you are not smart. On the other hand, people with a **growth mindset** think that intelligence is malleable. They know that it takes work to get smarter and that struggles and failures are not a sign of weakness, but a chance to learn. We'll talk a lot more about struggling and making mistakes later on. Carol Dweck, a psychology professor at Stanford University, coined these terms after years of research. The premise of her research is NOT that everyone has the same native intelligence or that intelligence isn't important, but instead that everyone can increase their intelligence.

Most people have a mixture of these two mindsets. In fact, it is possible to have a growth mindset about one thing (e.g. sports) while having a fixed mindset about another (e.g. math). To get some practice thinking about these two mindsets we will do an exercise.

Exercise 1.1 Imagine you are a freshman in college and you just completed your first test. You scored a 74, which is a lot worse than you would have liked. How would a person with a fixed mindset respond to this situation? Now consider how a person with a growth mindset would respond.

_____

_____

_____

_____

_____

_____

_____

_____

Exercise 1.2 Describe a time when you had a growth mindset and what the result was. It doesn't have to be about school; you can also describe your experience with sports, art, learning a new skill, *etc.*

_____

_____

_____

_____

_____

_____

_____

1.2 Struggling and making mistakes

One of the reasons that it was difficult to understand these concepts until recently is that there was not much research into the brain and how it works. A baby is born with most of his or her brain cells, called neurons. What allows a person to learn are the connections between these neurons. In order for learning to take place, the new information has to make a connection to something that the person already knows. We now know that brains can continue to grow and adapt as these connections occur. A common analogy used is to think of the brain as a muscle. If you want to build muscle, you need to consistently lift weights, and you can't just stick with a weight that feels comfortable. You need to constantly push yourself to lift heavier and heavier weights. If you want to learn, you need to not only put time in to studying but also put yourself in uncomfortable situations where you won't always know the answer.

Exercise 1.3 Do these addition problems.

3 + 6 =   5 + 2 =   6 + 8 =

Did you learn anything by doing these problems?

_____

_____

   Beth is a kindergarten student who is first learning addition. These problems may have been difficult for her and she *did* learn by doing them. But for you, addition is old hat and you probably have problems like this memorized because you have built automaticity. This information is already in your long term memory, so no learning took place by just recalling these facts. To figure out 3 + 6 Beth had to figure out which strategy to use. Maybe she decided to count on her fingers but realized that it would be quicker to start at the 6. So she held up 6 fingers and then counted 3 more to get 7, 8, 9. For you this problem was a simple memorization recall, but for her this problem was genuine problem solving. She was training her brain to make it grow. Often times this training is not comfortable. As you learn, if you are doing it correctly, you will make many mistakes and struggle often. The key thing to remember is that these moments of making mistakes and struggling are signs that you are learning.

Exercise 1.4 Write about a time when you struggled to learn something. What did it feel like?

_____

_____

_____

_____

Let's talk a little bit more about this idea of struggling. Sometimes struggling is pointless. Most people are not strong enough to lift a car. If I try to lift my car, I can struggle all I want, but it just isn't going to happen. The kind of struggle that *is* helpful is typically called "productive struggle." To keep with the lifting heavy objects analogy, let's say that I can currently bench press 100 pounds. If I go to the gym and lift 100 pounds I am practicing something I can already do so I will never get stronger. This is similar to Exercise 1.3 in that you did math problems you already knew how to do so no learning took place. On the other hand, if I try to lift 150 pounds this is probably pointless because it is too far out of my reach. If I try to lift 110 pounds that would be productive because it is a reasonable step from what I can already do. The same goes for math. You want to work on problems that are just a little bit hard for you so that you struggle, but if you work through that struggle then you will be able to find a solution. Of course, this is easier said than done. One way to help you understand what level you are at is through reflection. That is why you will see (and have already seen) many introspective exercises in this book. Another thing that will help you progress is setting goals. We will talk more about goals in Chapter 3.

As you work on problems that are just out of your reach, you will inevitably make mistakes and experience failures.

Exercise 1.5 Think of a very successful famous person (it could be an athlete, historical figure, actor, businessperson, *etc.*) and describe how you think this person achieved their success.

_____

_____

_____

_____

_____

You may think that famous people are lucky to have been born with a great talent or intelligence. What you can't see, however, is all of their failed attempts beforehand and all of their practice, persistence, and hard work. *New York Times* columnist Peter Sims describes this idea by saying

> "imperfection is a part of any creative process and of life, yet for some reason we live in a culture that has a paralyzing fear of failure, which prevents action and hardens a rigid perfectionism."

The following are examples of famous people who had many failures before achieving success. Many authors, such as J.K. Rowling, Stephen King, and even Dr. Seuss, got multiple rejections from publishers before finally getting published. These *New York Times* bestsellers were rejected sometimes as much as 30 times before their book was published. Thomas Edison is well known for his revolutionary inventions such as the light-bulb, bifocals, and the phonograph. Not as well-known are all of the times that he tried and failed. He is credited with saying,

> "I have not failed. I've just found 10,000 ways that won't work."

In computing there are many examples of people failing miserably before they became successful. After founding Apple, Steve Jobs got fired from his own company. After trying out different jobs he eventually was named Apple's CEO and helped morph Apple into the successful company it is today. Many athletes are also very persistent. Michael Jordan, in particular, is known for having a growth mindset. Here is a great quote from him:

> "I've missed more than 9,000 shots in my career. I've lost almost 300 games. 26 times, I've been trusted to take the game winning shot and missed. I've failed over and over and over again in my life. And that is why I succeed."

Babe Ruth once held the record for the most strikeouts. His attitude was

> "every strike brings me closer to the next home run."

That mindset helped him to become one of baseball's greatest players.

Exercise 1.6 How does making a mistake in math class make you feel? If it makes you feel embarrassed or uncomfortable, think of some ways to overcome this feeling. If you already view mistakes as an opportunity to learn then explain how you came to this point of view.

_____

_____

_____

_____

_____

_____

_____

_____

When you encounter a problem that you aren't able to do, it is psychologically helpful to use the word "yet." For example, if you are in this class then you probably don't know how to write a mathematical proof. Instead of saying either aloud or to yourself that you don't know how to write a proof, try saying "I don't know how to write a proof yet." Just adding this one little word gives you permission to not stress out if you don't understand something right away. It's a reminder that learning takes time and even though you don't have a concept down right now, it doesn't mean that you won't get there.

Exercise 1.7 Think of 3 things that you aren't able to do but would like to learn (they don't have to be school-related). Write a sentence for each one saying what you don't know how to do and then add the word "yet" at the end. Here's an example: I am not fluent in Spanish…yet.

_____

_____

_____

_____

1.3 Speed

Exercise 1.8 Pretend that you are learning math and your teacher asked a difficult question. You need to take your time and think about the question, but as you are thinking you see several other students raise their hand. You know that those students tend to answer questions correctly. Describe how you would feel in this situation.

_____

_____

_____

_____

_____

_____

It is a common misconception that in order to be smart you have to think fast. This is simply not true. There are many slow thinkers who are as smart, or even smarter than, fast thinkers. Here is a quote from a famous French mathematician named Laurent Schwartz.

I was always deeply uncertain about my own intellectual capacity; I thought I was unintelligent. And it is true that I was, and still am, rather slow. I need time to seize things because I always need to understand them fully...Towards the end of the eleventh grade, I secretly thought of myself as stupid. I worried about this for a long time. Not only did I believe I was stupid, but I couldn't understand the contradiction between this stupidity and my good grades. I never talked about this to anyone, but I always felt convinced that my imposture would someday be revealed: the whole world and myself would finally see that what looked like intelligence was really just an illusion. If this ever happened, apparently no one noticed it, and I'm still just as slow...At the end of the eleventh grade, I took the measure of the situation, and came to the conclusion that rapidity doesn't have a precise relation to intelligence. What is important is to deeply understand things and their relations to each other. This is where intelligence lies. The fact of being quick or slow isn't really relevant. Naturally, it's helpful to be quick, like it is to have a good memory. But it's neither necessary nor sufficient for intellectual success [2001, p. 30–31].

It is important to keep this in mind as you work on proofs. Remember that just because you didn't see the solution right away or it seems to take you a while to think about a good way to start a proof, that doesn't mean that you are bad at proofs. If a classmate is able to figure out a step faster than you that doesn't necessarily mean that person is smarter than you. It is perfectly okay to be a slow thinker.

1.4 What is math?

Exercise 1.9 How would you define mathematics?

Ask most people what they think mathematics is and they will say something about numbers, calculations, and following procedures. This is because throughout high school, math is usually taught as a set of rules and procedures where there is always a correct answer. Ask a mathematician what math is and they will speak about patterns, beauty, and creativity. Here is how the mathematician Keith Devlin describes math:

> "Mathematical thinking is a whole way of looking at things, of stripping them down to their numerical, structural, or logical essentials, and of analyzing the underlying patterns" [2011, p. 59].

As a math major in college you will begin to see this transformation from thinking of math as a set of rules to thinking of math as a creative subject as you begin to take upper level math courses. This introduction to mathematical proofs course is like a bridge between those two ways of viewing math. As a result, you may see a change in how you view yourself as a mathematician. This may be the first time that you won't be able to get the correct answer right away. You may have to put some time into thinking before starting a problem or you may try out a method that doesn't work and then have to start over. Now that we have learned about the growth mindset and the current brain research, you are better equipped to handle this. As we go through later chapters you will gain additional strategies in order to help you further.

Chapter 1 Homework

Pretend that you have a friend with a fixed mindset about mathematics. This friend thinks that being good at math comes naturally and therefore it should not require much effort. Write a letter to this friend sharing what you learned in this chapter to try to convince your friend to take the first steps to change their mindset.

_____

_____

_____

_____

_____

_____

_____

_____

_____

_____

_____

_____

_____

_____

_____

Growth Mindset Pledge

Every day I will do my best.
It's okay if I need to rest.

If things don't go my way,
I'll just try again another day.

If I make a mistake,
I will celebrate.

This chance to grow my brain.
And I will not complain.

I will persevere and muddle,
Through productive struggle.

And I will say,
Well, it's good to have a growth mindset anyway!

_____
Signature

Chapter 2

Team Building

2.1 History of cooperative learning

Mathematicians often collaborate, yet many classrooms still have students working individually in silence. A 2017 survey conducted by the National Association of Colleges and Employees found that employers rate teamwork and collaboration as top job skills that employees need [NACE, 2017]. This chapter describes the history and definition of cooperative learning. If you are not interested in background information and are anxious to see how cooperative learning can be used in the classroom, then skip ahead to the team building activities in Section 2.3.

Modern cooperative learning theory has its roots in the works of psychologists John Dewey, Jean Piaget, and Lev Vygotsky. In *The School and Society*, Dewey discussed his ideas on the deficient state of education and his plans on how it could be improved. Dewey defined a society as people working together because they share a common goal. In schools, he thought, this idea of joint effort was lacking. He believed that children automatically behaved like they were part of a society when they were on the playground, but in the classroom they failed to cooperate. The "tragic weakness" of schools, in his opinion, was that they tried to prepare future members of a society but failed to do so because they did not promote cooperation. Instead the focus is on

> "...the mere absorbing of facts and truths [which] is so exclusively individual an affair that it tends very naturally to pass into selfishness" [Dewey, 1896/1980, p. 10].

Students did not have a motive to become part of society when they studied. Their only motive was to compete with other students. In that type of atmosphere, cooperating became a negative goal. Dewey sought a new form of learning, a kind of learning in which students worked in groups.

Piaget was also interested in cooperative learning, but rather than focusing on the classroom as a society, he emphasized using peer experiences to help children learn. Piaget believed that peer interactions were crucial for a child's construction of social and moral feelings, values, and social and intellectual competence [1932/1965]. He defined cooperation as a social interaction between individuals who consider themselves to be equals and treat each other as such. He wrote that when students worked with their peers, they began to realize that their own understanding is personal and individualistic [1932/1965]. During cooperative learning, students heard how their peers interpreted and understood classroom material and this allowed for the development of alternative ways of comprehending the material. Piaget noted that schools demanded that students work in isolation. This went against a child's psychological tendencies, which are to work cooperatively.

> "Instead of taking into account the child's deeper psychological tendencies which urge him to work with others...our schools condemn the pupil to work in isolation" [Piaget, 1932/1965, p. 287].

Vygotsky [1978] created the social development theory. This theory states that social interaction plays a strong role in the development of cognition.

> "With assistance, every child can do more than he can by himself — though only within the limits set by the state of his development" [Vygotsky, 1978, p. 187].

Most investigations done to measure children's levels of development only took into account their individual activity. Vygotsky suggested that there is more to development than just the individual. He did experiments in which two children at the same development level worked together to solve problems. The children were able to perform better by collaborating than by working alone.

"What the child can do in cooperation today he can do alone tomorrow" [Vygotsky, 1978, p. 188].

In Vygotsky's experiments, interactions with others became internalized and transformed to produce new understanding that could be applied individually.

Exercise 2.1 Share an example from your own life where you saw Vygotsky's social development theory in action.

_____

_____

_____

_____

_____

_____

_____

_____

These three thinkers laid the foundation of cooperative learning theory. Dewey thought that learning together would help students become part of a functioning society. Piaget thought that learning together was natural for humans. Vygotsky believed that cognition was enhanced by cooperative learning. However, even though Dewey, Piaget, and Vygotsky all discussed cooperative learning, it was not the main focus of their works. Kurt Lewin, however, dedicated considerable effort to studying group dynamics and how they relate to social conflicts.

It was Lewin who developed a theory of how social cooperation takes place, and it would be his intellectual descendants who would apply that theory to education. The two key ideas that emerged out of Lewin's work were interdependence of fate and task interdependence.

"Interdependence of fate" is defined as occurring when individuals realized that their own fate was dependent on the fate of the group. It is only when fate was shared that the collection of individuals could be called a group [Lewin, 1935]. Task interdependence occurred when group members depended on each other to achieve their goal. Lewin argued that people come to a group with different dispositions, but if they share a common objective, they are more likely to act together to achieve it [Smith, 2001]. Lewin also developed a theory of motivation which states that a drive for goal accomplishment is what motivates a person towards either cooperative, competitive, or individualistic behavior.

Exercise 2.2 Why are interdependence of fate and task interdependence important components of cooperative group work?

_____

_____

_____

_____

_____

_____

_____

_____

Morton Deutsch was one of Lewin's students. Deutsch built upon Lewin's cooperative group theories. Deutsch [1949] developed Lewin's theory of motivation into a theory of cooperative and competitive environments. Cooperation occurred when group members had positive interdependence, meaning that the goals of each individual correlated positively with the goals of other members of the group. Competition occurred when group members had negative interdependence, meaning that the goals of each individual had a negative correlation with the goals of other group members. Deutsch found that students working in a cooperative environment showed more acceptance of others' ideas, more combined efforts, more division of labor, and not as many communication problems as students working in a competitive environment.

Exercise 2.3 As a student, how would you feel in a competitive class-room? How would you feel in a cooperative classroom?

_____

_____

_____

_____

_____

_____

_____

David Johnson was one of Deutsch's students. Johnson, along with his brother Roger, built on Deutsch's theories about group dynamics and transferred them into the classroom. The Johnsons have become prom-inent figures in the field of cooperative learning along with Robert Slavin and Spencer Kagan. It was these people who defined the contemporary field of cooperative learning.

2.2 Components of cooperative learning

According to cooperative learning theorists, the defining characteristics of cooperative learning are: positive interdependence, individual account-ability, peer assistance, equal opportunity, heterogeneous grouping, close-ness, and processing.

"Positive interdependence" is the term used to describe the fact that each member of a group is dependent on other members of the group for accomplishing the educational goal and for learning. You may have heard the phrase "we sink or swim together." That phrase really gets at the idea of positive interdependence. It provides several advantages. Crucially, students themselves can discern the value of positive inter-dependence because they understand that they are linked with other members of the group in a way such that they cannot succeed unless the other members do [Johnson & Johnson, 1994]. Another advantage is that this positive interdependence makes each student feel responsible for the learning of all other group members.

Exercise 2.4 Describe ways that your teacher can promote positive interdependence when you work in groups during class.

"Individual accountability" means that each student is responsible for learning the material. Individual accountability focuses the activity so that group members are explaining concepts to one another [Slavin, 1991]. It is individual accountability that allows students to make sure that each group member learns the material. Making sure that equal opportunities exist will then help to solidify positive interdependence. Students learn best when they all participate about equally [Kagan, 2001].

Individual accountability can be obtained by keeping the group size small (usually 3 or 4 per group), assessing students individually, randomly calling on a student to present the group's work, and by observing the groups to make sure that each student is participating. Individual accountability is a crucial element of cooperative learning because without it problems such as social loafing or division of labor can occur. Social loafing occurs when one student in the group does relatively little work with the expectation that the other group members will get the task done. Social loafing may not hinder group productivity, but it does hurt individual learning. Division of labor occurs when the group members decide to divide up the work. If this happens, each student is not getting the full learning experience, and just like with loafing, this will hinder individual learning.

"Peer assistance" is informally defined as students helping each other. Without peer assistance, students would be sitting together in groups but they would not be working together cooperatively. While the students are assisting each other they engage in discussions by helping and encouraging each other. Peer assistance occurs when individuals encourage and facilitate each other's efforts to complete tasks in order to achieve the group's goal. It is this peer assistance that allows the learning to become student-centered.

Exercise 2.5 Discuss a time as a student when your group didn't have peer assistance. The group was physically sitting together but no peer assistance was going on. Was this helpful?

_____

_____

_____

_____

_____

_____

_____

"Equal opportunity" means that each student has an equal chance of participating in the group activity and, in turn, learning the content. Equal opportunity allows everybody to be equally challenged and to make sure that everybody contributes. Making sure that equal opportunities exist will then help to solidify positive interdependence. Students learn best when they all participate about equally.

It is important for groups to be heterogeneous in terms of ability level, gender, and ethnicity. When groups are maximally heterogeneous, students tend to become tolerant of diverse viewpoints [Stahl, 1994]. Heterogeneous grouping is also useful as it encourages different perspectives. It has been found that when students are in groups where members are of the same ability level they ask each other questions but receive no response almost four times as frequently as when they are in mixed ability groups [Ashman and Gillies, 1997]. When groups are formed randomly they tend to be heterogeneous groups.

Physical closeness is another key idea in cooperative learning. Students should be close enough to touch each other without actually touching. If the students are close to each other they will be more likely to feel like a group. Proximity makes a difference in how the students act and interact with each other.

To achieve all of the components of cooperative learning, it is important for groups to have time to practice working together. It is easy to sit together with other people, but that doesn't guarantee that you will be working in a cooperative group. For the remaining exercises in this chapter you should be sitting in a group. You want to achieve closeness so make sure that your desks are arranged as shown in the picture.

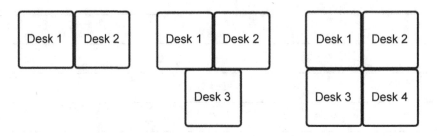

Figure 2.1. Desk placement

If you are sitting at round tables then you are already in a good position. If you are sitting at long rectangular tables, sit so that two people are next to each other on one side and the remaining one or two people are sitting on the opposite side.

2.3 Group tasks

Exercise 2.6 For the first task, discuss with your group good and bad behaviors that might occur while working together in a math class. What does a group that works well together look and sound like? What does a group that works together poorly look and sound like?

Table 2.1. Good and bad group work behaviors

Works Well	Works Poorly

After your group has a chance to brainstorm, have a class discussion where each group can share some of their ideas. The reason why solving a mathematical proof lends itself to group work is because group-worthy tasks are open-ended. This means that there is more than one way to arrive at a solution and there can be more than one correct answer, much like the task you just completed. Solving these types of problems will lead to group discussions and to more flexible thinking strategies. Tasks should also be productively uncertain, where you do not know how to get to the "answer" the teacher is looking for and therefore must perform complex problem solving. The problem solving encourages you to debate and work through issues together and learn from your mistakes.

For the next task we will work on developing positive interdependence. This task is taken from a blog post by Sarah VanDerWerf [2015].

Exercise 2.7 This is called the *100#* task. Each group gets a highlighter and one copy of the activity sheet (see Appendix A). You will get 3 minutes to highlight as many numbers as you can going in order from 1 to 100. Group members should take turns highlighting their number. The first person highlights number 1 then passes the sheet and highlighter to person 2, who has to highlight number 2, and so on. You can help each other out, but do not go out of order. After the first round discuss what went well, what could go better, and come up with a strategy for the second round.

Round 1 score _____. What is your strategy for round 2?

_____

_____

_____

_____

Now your group will receive another copy of the activity sheet. Again, you will get 3 minutes to repeat this activity. Try to beat your round 1 score.

Round 2 score _____. Did your strategy work? What patterns did your group notice?

_____

_____

_____

_____

Did you feel a sense of positive interdependence (we are all in this together)?

_____

_____

_____

_____

In this next task we will work on preventing dominance. You will need to develop the ability to gather information from each of your group's members, make a plan together, agree on strategies, and share ideas. There are many different ways to accomplish this task so you will have to give reasons for your ideas to try to convince your group that you are correct. You will, in turn, need to listen to the ideas of other group members.

Exercise 2.8 You decide to go on a hike in the wilderness. You only have room to bring 10 items. You are asked to rank these items in order of importance from the most important (1) to the least important (10). In the first stage you will work by yourself to come up with a ranking. After that you will work with your group to reach a consensus on a ranking. Once all groups are finished, you will be shown an experts' ranking. Next, calculate the difference between your individual rankings and the expert rankings, and then your group rankings against the expert rankings. The following table can be used to generate a comparison. Use the table's columns to measure how far away you are from the expert ranking, regardless of direction. See if you did better while working alone or with your group.

Table 2.2. Hiking activity

Item	Individual	Group	Expert	$|I - E|$	$|G - E|$
Food					
Headlamp with batteries					
First aid kit					
Extra clothes					
Lighter					
Water					
Compass					
Shelter					
Sunscreen					
Knife					
			Total		

Exercise 2.9 How successful were you at working together, sharing ideas, and listening to other people's ideas?

_____

_____

_____

_____

_____

_____

Chapter 3

Setting Goals

3.1 SMART goals

The first step to setting goals is clarity. In other words, you need to be clear on what your goal is. To help with this step we will look at SMART goals. SMART stands for specific, measurable, attainable, relevant, and timely.

Your goal should be specific because a vague goal is hard to achieve. You want to be clear about what you want to accomplish so you will know how to go about achieving it. Think about what you want to accomplish and why it is important.

Your goal should be measurable because you need to know when it is achieved. A goal such as "I want to make a lot of money" is not measurable because it is unclear what "a lot" means. There is no measure for this and no way to know if this goal is achieved.

You want your goal to be attainable. This is similar to what was discussed in Chapter 1 about productive struggle. We said you want to solve math problems that are just beyond your reach so that learning occurs. The same idea applies with setting goals; you want to set a goal that is possible, but at the same time not too easy. A goal that is too easy will not be motivating. A goal that is too difficult will be discouraging. For example, a goal to live on Mars is not attainable at this point in time so if this is your goal it will be hard to make any progress. An important thing to remember is that your goal must be something that is within your control. A typical first attempt at an educational goal is to get an A in a course. This is not within your control; your professor is the one who assigns the grade. You want your goal to be process-based rather than outcome-based. Getting an A is the outcome but think about the process of how you would get that A.

In terms of relevancy, you want to make sure that your goal is something that you care about achieving. If a goal is not important to you then you won't be motivated to achieve it.

Being time-bound is important because it creates a sense of urgency. Without a time limit your goal may become something that you want to achieve "someday." Pretty soon someday turns into never. Think about something that you one day hope to achieve but never set a time limit on. You probably have never put effort into achieving it because there is no pressure to get started. For example, many people say they would one day like to learn a foreign language. Unless they specifically say they would like to learn it, say, within a year, it is unlikely that they will make any progress. As the governor of Rhode Island Gina Raimondo said, "the difference between a dream and a goal is a deadline."

Let's go through an example of how the thought process could work when setting a SMART goal. You may want to start by thinking about the outcome you wish to achieve. We will use the example of getting an A in this course. Next, think about different ways you can achieve this. Some ways could be by always attending class, doing every homework, planning to go to office hours, forming a study group, getting enough sleep the night before an exam, and setting aside an hour every day to do math problems. All of these are within your control, are specific, and are attainable. Some of these are already measurable and timely. For those that aren't, you can add a time component. For example, if you decide to form a study group then your goal could be to form a study group the first week of school and plan to meet twice a week throughout the semester for at least one hour. Academic goals don't always have to be tied to grades. You may choose other outcomes, such as know how to write a mathematical proof or be neat and organized with your work. If you look at your syllabus, most likely your professor has listed course goals. These are what your professor thinks are the most important outcomes of taking this course.

Exercise 3.1 Pick one long term goal for this class over the whole semester and write it as a SMART goal. Make sure to include all of the components discussed above. You can start with the desired outcome and then answer the following questions to make sure it is SMART: What specifically do you want to happen? What steps will you take to achieve your goal? How will you know when your goal is achieved?

_____

_____

_____

_____

_____

_____

_____

_____

3.2 Motivation

The second step to setting goals is motivation. This step is crucial because you may have a really great goal, but without motivation you will never achieve it. The desire to achieve a certain grade may at first glance seem to be motivating. In fact, it is beneficial to think beyond grades because studies have shown that grades are not motivating and instead they enhance anxiety and cause students to avoid challenging courses [Chamberlin, Yasue, and Chiang, 2018]. Another type of motivation is monetary. An example of a company that utilizes this type of motivation is HealthyWage. They allow someone to place a bet on losing weight by setting a weight loss target and amount to bet each month. If that person wins the bet, they get a pre-determined cash prize. Otherwise, they lose the money that was bet. Think about if this would be helpful for you to achieve your goals. You can make a bet with a friend or family member. HealthyWage also allows people to form teams that compete for weight loss. This ties back to cooperative learning and the idea of positive interdependence; in order for one group member to succeed they all must succeed. That can be a form of motivation because being part of a group makes you feel responsible for other group members. One way to achieve this in the classroom can be to set group goals rather than individual goals. Working in a group and sharing goals with your group can also help with the last step for setting goals, which is accountability.

Exercise 3.2 List some possible motivators that you think will help you achieve your goal.

_____

_____

_____

_____

_____

For some people accountability may just mean self-accountability. Nutritionists often suggest that their clients keep a food diary by writing down everything that they eat. Often the nutritionist never looks through this diary, but the act of simply writing down all foods eaten adds an element of self-accountability that, for some people, makes them eat healthier. Writing your goals down may be enough accountability for you. Other people need outside accountability. This can be a friend in class to whom you report goal progress. Or you can choose to document your goal progress on social media and have your friends there hold you accountable.

Exercise 3.3 What method of accountability do you think will work best for you? Describe your plan here.

_____

_____

_____

_____

_____

Chapter 4

Logic

4.1 Statements

A statement is a sentence that can only have a true or false value. An example of a statement is: the floor is blue. You can look at the floor and see whether it is blue or not. If the floor is blue then this is a true statement. If the floor is not blue then this is a false statement. What would make a sentence not a statement? There are several possibilities. It could instead be a question or an exclamation. It could also be too ambiguous to be a statement. For example, consider the sentence "she went to the store." In that sentence it is unclear who "she" is referring to and depending on who it refers to, that sentence could either be true or false. Later in this chapter we will see a way to change this ambiguous type of sentence into a statement. For now, we will focus on ways to combine statements to form new statements.

The first way to combine two statements is by putting an "and" between them. For example, the floor is blue and the ceiling is yellow. The symbol used for "and" is like a capital A without the center line, written ∧. You can also put an "or" between statements. This is represented by the symbol ∨.

The next way to combine statements is called a conditional statement. This is a statement that has an if-then in it. For example, if the floor is blue, then the desk is yellow. The notation used to represent a conditional statement is an arrow. We can let p stand for "the floor is blue" and q stand for "the desk is yellow." Then this would be written as $p \rightarrow q$ and read as "if p then q" or "p implies q." A biconditional statement is a statement that is conditional in both directions. An example is "today is Monday if and only if I play tennis." The "if and only if" in between each statement is sometimes abbreviated *iff*. We could write this biconditional statement as $p \leftrightarrow q$. It is equivalent to: if today is Monday then I play tennis and if I play tennis then today is Monday.

Exercise 4.1 Give an example of a biconditional statement and then write its equivalent form with two conditional statements.

_____

_____

_____

_____

Now we will look at a way to change a conditional statement by taking what is called a contrapositive. If the original statement is "if p then q," then its contrapositive will switch the order and negate each part, turning it into "if not q then not p." We use the "~" to symbolize the negation of a statement. Thus, the contrapositive can be written as $\sim q \rightarrow \sim p$. The contrapositive of the conditional statement "if today is Monday then I play tennis" is "if I don't play tennis then today is not Monday." One thing you want to notice is that the original conditional statement and its contrapositive have an equivalent truth value.

Exercise 4.2 Make up a conditional statement and write its contrapositive.

_____

_____

_____

_____

4.2 Number sets

Before we can move on to the next logic topic, we will take a break to review number sets. The first number set is the natural numbers, denoted $\mathbb{N}$. These are sometimes called the counting numbers. They are whole numbers beginning with 1. Building on the natural numbers, if you include 0 and the negative whole numbers then these are called integers. They are denoted by $\mathbb{Z}$ because the German word for number is zahlen. The next set is the set of rational numbers. These are denoted by $\mathbb{Q}$, where the q stands for quotient. This is because rational numbers are any number that can be written as a quotient, or fraction. To be precise, the term fraction means an integer divided by a non-zero integer. If you prefer to think in terms of decimals, a rational number is any terminating decimal or a non-terminating decimal that has a pattern to it.

Exercise 4.3 Give two examples of numbers with non-terminal decimals that are rational. Show this is true by expressing them as fractions.

_____

_____

_____

_____

Exercise 4.4 Explain why any integer is also a rational number.

_____

_____

_____

_____

The last number set we will discuss in this class is the set of real numbers. They are denoted by $\mathbb{R}$. These include all of the rational numbers and all of the irrational numbers. Irrationals can simply be defined as the opposite of rationals, meaning they cannot be expressed as a fraction. In terms of decimals, irrational numbers are non-terminating and do not have any kind of repeating pattern. The most famous irrational number is pi. In other classes you may also discuss complex numbers, which include real numbers as well as square roots of negative numbers.

Exercise 4.5 Look at the four choices and decide which one doesn't belong. Make sure you have a reason to back up your choice.

Table 4.1. Which one doesn't belong

.83791...	1/5
6	$.\overline{27}$

I think _____ doesn't belong because

4.3 Quantifiers

A quantifier gives quantity to a sentence. There are two types of quantifiers: the universal quantifier and the existential quantifier. The universal quantifier is written like ∀ and is read as "for all" or "for every." The existential quantifier is written like ∃ and is read as "there exits" or "there is." Earlier we had an example of a sentence that was not a statement because it was too ambiguous. We can use a quantifier to turn it into a statement. The original sentence was "she went to the store." Think of the word "she" as a variable, maybe call it x, that you can plug in different values for. Instead of using p like we usually do to represent a statement, we will use p(x) to represent a sentence with the variable x that can become a statement when a specific value of x is plugged in. This is called a predicate. The x represents a person, so we will say x comes from the set of people or x is an element of the set of people. This can be written as x ∈ {people}. The general format for using a universal quantifier is: ∀ x ∈ {domain}, p(x). For our example this would be written as ∀ x ∈ {people}, x went to the store. In common language this would be read as "all people went to the store." Think of plugging in the "all people" into where the x is.

Exercise 4.6 Write the following in formal notation with a quantifier: All dogs eat food.

The general format for using an existential quantifier is: $\exists\, x \in \{domain\}$ such that p(x). For our example this would be written as $\exists\, x \in \{people\}$ such that x went to the store. In common language this would be read as "some people went to the store."

Exercise 4.7 Write the following in formal notation with a quantifier: Some cats are friendly.

Exercise 4.8 Write two statements, one with each quantifier. For each write it in the formal notation and the informal notation. When you write the informal notation be careful to put it in the kind of language somebody would use in everyday speech rather than just translating the formal notation.

The kind of statements you wrote in the last exercise are in the form that mathematical statements to prove are often written in. Usually the proof is for all values because it is more useful to prove something in the general case than to have to prove multiple, usually infinite, specific cases. Statements to prove are also usually written as conditional statements because some information will be known, whatever is written between the "if" and the "then," and the part after the "then" is what you want to prove. We learned about the contrapositive of a conditional statement because it will also be useful when we learn a specific proof technique called proof by contraposition. The next logic topic we will look at will also be useful with a specific proof technique.

4.4 Negations

Sometimes it will be helpful to prove a statement by contradiction. For this technique we need to know how to take the negation of various types of statements. To think about how negations work we will use a short book with pictures. This book is titled *Frank the Fox is a Liar!* You can view the book online at https://tinyurl.com/frankthefox.

On Monday Frank said, "my friend Cassie the cat has blue fur and she is a cat." Cassie is a cat, this much is true. But Cassie was very proud of her grey fur. Frank lied on Monday.

Figure 4.1. *Frank the Fox* Page 1

Exercise 4.9 How do you know Frank lied on Monday? In general, what would make an "and" statement false? What is the negation of p and q?

_____

_____

_____

_____

_____

On Tuesday Frank saw Bella the bird. He said, "Bella does not have feathers or she is not flying." Oh Frank, you are still a liar!

Figure 4.2. *Frank the Fox* Page 2

Exercise 4.10 How do you know Frank lied on Tuesday? In general, what would make an "or" statement false? What is the negation of p or q?

_____

_____

_____

_____

_____

_____

_____

_____

On Wednesday Frank saw his friend Floe the flamingo. He said, "If today is Wednesday then I will not see any flamingos." Frank has lied once again.

Figure 4.3. *Frank the Fox* Page 3

Exercise 4.11 How do you know Frank lied on Wednesday? In general, what would make a conditional statement false? What is the negation of if p then q?

_____

_____

_____

_____

_____

_____

_____

_____

_____

On Thursday Frank picked some flowers and he said, "all of the flowers are the same." Another lie.

Figure 4.4. *Frank the Fox* Page 4

Exercise 4.12 How do you know Frank lied on Thursday? In general, what would make a "for all" statement false? What is the negation of ∀ x ∈ {domain}, p(x)?

_____

_____

_____

_____

_____

_____

_____

_____

_____

On Friday Frank wanted to eat a snack and
he said, "some of my bananas are opened."
No surprise, Frank was lying again.

Figure 4.5.   *Frank the Fox* Page 5

Exercise 4.13 How do you know Frank lied on Friday? In general, what would make a "there exists" statement false? What is the negation of $\exists\, x \in \{\text{domain}\}$ such that p(x)?

_____

_____

_____

_____

_____

_____

_____

_____

_____

Exercise 4.14 Write three statements that contain a mixture of all of the types of statements discussed and then find the negation of each.

_____

_____

_____

_____

_____

_____

4.5 Logical arguments

The last logic topic we will look at is logical arguments. These will help you begin to think about how mathematical proofs are structured. A logical argument has a list of givens, called premises, followed by a conclusion. If the argument is valid, then you would be able to deduce the conclusion from the premises. For this topic it will be useful to build a collection of popular valid arguments. The first example we will look at is called Modus Ponens. This argument starts with two premises. The first is the conditional statement "if p is true then q is true." The other is "p is true." With these two given statements it would make sense that q would also be true. Written out, Modus Ponens looks like:

$p \rightarrow q$

p

$\therefore q$

The symbol with three dots can be read as "therefore." Following this symbol is the conclusion. The next argument is similar. Earlier we said that a conditional statement and its contrapositive are equivalent. The first given is if p then q, but we will replace this with the contrapositive, $\sim q \rightarrow \sim p$. Then follow the same argument as Modus Ponens. The next given is $\sim q$. From those we can conclude $\sim p$. This argument is called Modus Tollens.

Exercise 4.15 Write an argument in words that follows either Modus Ponens or Modus Tollens.

_____

_____

_____

_____

_____

_____

The next argument shows how we can build an "or" statement. It is called generalization. If you know a statement is true then you can conclude that an "or" statement that includes that statement will be true. Here is an example.

Today is Monday.
Therefore, today is Monday or my skin is purple.

Exercise 4.16 Write the general format for generalization.

_____

_____

_____

_____

For an "and" statement to be true you need both parts to be true. Using this idea, we can build two different logical arguments using "and" statements. The first is called specialization. Here the premise is just a single "and" statement. From that you can conclude either of the individual statements. The other argument is called conjunction. In this argument the premises are two different statements and from that you can conclude the statement that joins them together with an "and" is true.

Exercise 4.17 Give an example in words of either specialization or conjunction.

The next argument is called elimination. In this argument we are given p or q is true as well as ~p is true.

Exercise 4.18 What can you conclude?

The next argument is called transitivity. You may have heard this term before. For example, the equals sign is transitive; if a = b and b = c then a = c. It turns out that conditional statements are also transitive; if p → q and q → r then we can conclude p → r.

The last argument we will look at is called proof by division into cases. In this argument we know that p or q is true. We also know that p → r and q → r. From these premises we can conclude that r must be true. Here is a table showing all of the above mentioned valid arguments.

Table 4.2. Logical arguments

Modus Ponens	Modus Tollens
$p \rightarrow q$ p $\therefore q$	$p \rightarrow q$ $\sim q$ $\therefore \sim p$
Generalization	Specialization
p $\therefore p \vee q$	$p \wedge q$ $\therefore p$
Conjunction	Elimination
p q $\therefore p \wedge q$	$p \vee q$ $\sim p$ $\therefore q$
Transitivity	Proof by Division into Cases
$p \rightarrow q$ $q \rightarrow r$ $\therefore p \rightarrow r$	$p \vee q$ $p \rightarrow r$ $q \rightarrow r$ $\therefore r$

From these known arguments we will be able to show that other arguments are also valid. We will start with an example showing that the following argument is valid.

$p \rightarrow a$

$p \rightarrow b$

p

$\therefore a \wedge b$

Usually a good place to start if you see a single statement is to try to pair it with another statement to try to mimic one of our known arguments. In this case the p is by itself. We can use it with either of the first two conditional statements. Given the premises $p \rightarrow a$ and p we can conclude a by Modus Ponens. Similarly, using the premises $p \rightarrow b$ and p we can conclude b also by Modus Ponens. Once you used a known argument to reach a conclusion you can think of adding that result to your givens. We now know that a and b are both true. Then, by conjunction, we can conclude a and b is true. This is what we were trying to show.

Exercise 4.19 Describe the process we went through to show an argument is valid.

_____

_____

_____

Exercise 4.20 Show that the following arguments are valid.

$s \rightarrow r$	$(\sim p \vee q) \rightarrow r$
$(p \vee q) \rightarrow \sim r$	$s \vee \sim q$
$\sim s \rightarrow (\sim q \rightarrow r)$	$\sim t$
p	$p \rightarrow t$
$\therefore q$	$(\sim p \wedge r) \rightarrow \sim s$
	$\therefore \sim q$

Chapter 4 Homework

(1) Explain the difference between a conditional and a biconditional statement. Give an example of each.

(2) Write two examples of statements and their contrapositive. Explain why the original statement and its contrapositive are equivalent.

(3) Write the negation of each statement:
    (a) The lights are on if and only if Bob is at home.
    (b) $\forall$ integers x, if $x \neq 0$ and $x \neq 1$ then $x^2 > x$.
    (c) Some desks are purple or all chairs are blue.
    (d) If it doesn't rain, then the grass will not grow.
    (e) If we do not go to school on Memorial Day then Memorial Day is a holiday or we do not work on Memorial Day.

(4) Show the following argument is valid.

$$(p \vee r) \to (s \wedge t)$$
$$p$$
$$\therefore t$$

(5) Show the following argument is valid.

$$p \to q$$
$$r \vee s$$
$$\sim s \to \sim t$$
$$\sim q \vee s$$
$$\sim s$$
$$(\sim p \wedge r) \to u$$
$$w \vee t$$
$$\therefore u \wedge w$$

Chapter 5

Problem Solving

5.1 Metacognition

Before we dive into problem solving it will be useful to discuss meta-cognition. Metacognition is thinking about thinking. That definition might sound funny, but the idea is that it is useful to understand and reflect on how you think. To illustrate this, we will go through an example.

Exercise 5.1 With a partner play two games of tic-tac-toe. On the first game, write down your thought process for each move. You don't need to write in full sentences. On the second game both you and your partner speak your thoughts out loud.

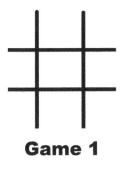

Game 1

Figure 5.1. Tic Tac Toe game 1

Thought Process

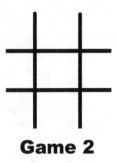

Game 2

Figure 5.2. Tic Tac Toe game 2

How were these games different from a regular game when you didn't take the time to think about your thoughts?

_____

_____

_____

_____

5.2   Creativity

Exercise 5.2 What does it mean for a person to be creative? Do you consider yourself to be creative? Why or why not?

_____

_____

_____

_____

_____

Many people think creativity is tied to the arts. In the following quote Sir Ken Robinson discusses creativity in his book *Creative Schools*.

> "Being creative is not just about having off-the-wall ideas and letting your imagination run free. It may involve all of that, but it also involves refining, testing, and focusing what you're doing. It's about original thinking on the part of the individual, and it's also about judging critically whether the work in process is taking the right shape and is worthwhile, at least for the person producing it" [2016, p. 119].

So while your initial thoughts about creativity may have led you to think of dancing or poetry, this description can also be used for doing mathematics. In this chapter we are going to start problem solving. Keep this quote in mind as you go through the problems. You will be making conjectures, refining them, testing them out through various techniques such as making visuals, and thinking creatively.

5.3 Problem solving

Probably the most famous problem solving structure was developed by George Polya in his book *How to Solve It* [Polya, 1957]. The four steps he gives for problem solving are: understand the problem, devise a plan, carry out the plan, and look back. We will unpack these steps while going through an example problem.

Example 5.1 Every birthday my cake has exactly as many candles as is my age that year. So far I have blown out 210 candles. How old am I?

The first step is to understand the problem. In this step make sure to read the problem carefully and to understand all of the terms. It is very tempting to dive right into solving the problem, but it is important to take a breath first and clear up any possible confusion with what the question is asking. A novice mistake is to answer the wrong question. In this step you may want to break the problem into parts. Think about if you have all of the tools that you need in order to solve it; is there enough information given? Is it at an appropriate skill level for you to be able to solve? In this step also begin to think about if you want to draw a picture to represent the situation. Visuals can be very useful and we often forget them as a possible problem solving tool. Think about if you want to introduce any variables or other notation. For this birthday problem it is clear that on my first birthday I had 1 candle, 2 on my second birthday, and so on. On a specific year I will have to add all the previous years' worth of birthday candles to get the total.

The second step is to devise a plan. This is your chance to explore. Don't be afraid to try ideas out. Keep in mind that not every idea you try will lead you to the solution, but everything you try will be useful in some way, even if just to rule out that method as a way to find the solution. Here are some suggestions for possible plans to try: guess and check, look for a pattern, draw a picture, eliminate possibilities, solve a simpler problem, consider special cases, and work backwards. You may end up using more than one of these. For this birthday problem I will make a visual and look for a pattern because I know that the early years will be easy to figure out and that once I have the answer for a year then the next year's answer will be easy to get because I am just adding my new age. You may have chosen a different plan. That's okay. The beauty of problem solving, and mathematics in general, is that there are many ways to get to the correct solution to a problem. It will also depend on your skill level. This problem could be given to an elementary student, who may choose to solve it by guess and check. The way the problem is stated is not difficult to understand and, depending on your skill level, you may be able to use a more sophisticated method for solving it.

The third step is to carry out the plan. This step is pretty clear; actually do what you said you planned to do. Sometimes you might have no idea how to get started. If this is the case then **just do something**. Draw a picture, plug a number in, or try to make the problem simpler. This step is where all of your training in growth mindset will come in handy. You already know that you learn from mistakes and that productive struggle is a good thing, an opportunity to grow your brain. I began to calculate the beginning terms to look for a pattern.

$$1$$
$$1 + 2 = 3$$
$$1 + 2 + 3 = 6$$
$$1 + 2 + 3 + 4 = 10$$

Figure 5.3. Birthday candle problem

From here I realized that I wanted to figure out when

$$1+2+3+\ldots+n = 210. \tag{5.1}$$

There is a nice formula for the sum of the first n terms that can be used to reduce the left side,

$$1+2+3+\ldots+n = \frac{n(n+1)}{2}. \tag{5.2}$$

We will actually prove this later on once we get to induction, but for now we can look at a method supposedly used by a young mathematician named Carl Friedrich Gauss. The story goes that Gauss was in his classroom and his teacher wanted to give busy work so she asked the class to add the first one hundred numbers. Gauss listed the numbers 1 to 100 in a row and then listed the same numbers but counting backwards on the row below. He then noticed that each pair summed to 101. There were 100 of them, for a total of 100 (101), but each number was counted twice. Thus, the formula he used was

$$\frac{100(101)}{2} = 5050. \tag{5.3}$$

Getting back to our problem, I now had to solve

$$\frac{n(n+1)}{2} = 210. \tag{5.4}$$

Multiplying both sides by 2 yields $n(n+1) = 420$. I could have solved this by trying to factor or use the quadratic formula, but instead I thought it would be quicker to look at $\sqrt{420} \approx 20.49$ because $n(n+1)$ is pretty close to a square number. This led me to believe that 20 times 21 should equal 420.

The last step is to look back. It is easy to solve the problem, feel triumphant, and want to stop there. This last step is important though because you want to make sure that your solution is correct and to reflect back on your method and solution. Is your answer reasonable? Did you actually answer the question? What did you learn? Will this help you solve other problems? I ended up getting that I am 20 years old. I can check this answer in multiple ways. One way is to check that

$$\frac{20(21)}{2} = 210. \tag{5.5}$$

It does! This answer seems to make sense and we now know a formula for summing the first n natural numbers that may be useful in later problems.

Exercise 5.3 Use Polya's method to solve the following problem: Find five positive whole numbers whose sum is 25 and product is 945.

Understand the problem:

_____

_____

_____

_____

Devise a plan:

_____

_____

_____

_____

Carry out the plan:

Look back:

Next we will go through several more examples to show how some of the other plans mentioned can be useful. One famous problem is called the handshake problem. It goes like this:

There are nine people in a room and every person shakes hands exactly once with all of the other people. How many handshakes will there be?

First, read through this problem and make sure that you understand it. Trying to solve this problem with nine people can be overwhelming. Instead, we will make the problem simpler and start with three people. Not only does it help to make the problem smaller, but you can also act out the problem to help solve it.

Exercise 5.4 How many handshakes will there be in a group of 3 people? What about a group of 4 people? Keep increasing the number of people until you notice a pattern.

Another famous problem is a variation of the following:

Lily pads grow in a pond and double their area every 24 hours. On the first day of spring there is only one lily pad on the water. Sixty days later the pond is completely covered. When is the pond half covered?

Exercise 5.5 Try solving this problem.

The simplest way to solve this problem is to use the working backwards technique. If the pond is completely covered on day 60 then on day 59 it must have been half covered.

Exercise 5.6 Solve the following problem using Polya's framework: You are on an island where each inhabitant is either a *truth-teller* or a *liar*. Truth-tellers *always* tell the truth and liars *always* lie. Mary and Steve are on the island.

Mary says: "If 28 is even then I am a truth-teller."
Steve says: "Mary is a liar."
Determine whether each person is a truth-teller or a liar.

Exercise 5.7 Solve the following problem using Polya's framework: Why does every year have to contain a Friday the 13th and what is the greatest number of Friday the 13th occurrences in a single year? [taken from: https://nrich.maths.org/610]

Exercise 5.8 Cryptarithms are puzzles that have non-numeric values in for calculations. Solve this famous cryptogram proposed by H.E. Dudeney in 1924. Each letter stands for a different digit and none of the numbers begins with zero. Figure out what each letter stands for.

```
  SEND
+ MORE
-------------
 MONEY
```

5.4 Games

A fun way to practice problem solving is through playing games. In this first game I am going to give you items that I like and items that I don't like. Try to figure out a way to know what I like and dislike from these examples.

I like balls, but I don't like bats.
I like yellow, but I don't like orange.
I like being silly, but I don't like laughing.
I like feet, but I don't like toes.
I like pools, but I don't like bathtubs.

When you think you have figured out the rule, don't tell anyone the rule. Instead, give a sentence of something I would like and dislike. I will tell you if I agree or disagree.

This next game is a variation of a game called Nim. There are 21 objects. You can choose to take one, two, or three of those objects on your turn. Your partner can do the same. The person who ends up taking the last object wins the game. You can see this game played out in an episode of *Survivor: Thailand* [2002, episode 6]. Here is an example of how the game might go:

> Player 1: I will take 3 items leaving 18 items.
> Player 2: I will take 2 items leaving 16 items.
> Player 1: I will take 3 items leaving 13 items.
> Player 2: I will take 1 item leaving 12 items.
> Player 1: I will take 3 items leaving 9 items.
> Player 2: I will take 1 item leaving 8 items.
> Player 1: I will take 1 item leaving 7 items.
> Player 2: I will take 3 items leaving 4 items.
> Player 1: I will take 3 items leaving 1 item.
> Player 2: I will take the last item and win!

Exercise 5.9 Play this game with a partner. Try to come up with a strategy to always win.

Our last game is called 101 and Out. This game comes from math educator Marilyn Burns [2007]. The goal of this game is to get as close to 101 points as possible without going over. The class will be divided in half and each group will get a die. Groups then take turns rolling the die and deciding to either count the number at face value or multiply it by 10. For example, if one group rolls a five they can keep that number or turn it into 50.

Exercise 5.10 Play this game with your class or with a partner. Discuss what happened when you played and what strategies you used.

_____

_____

_____

_____

_____

_____

_____

Chapter 5 Homework

Solve the following problems using Polya's method. Write out all the steps and any thoughts you have, even if they did not lead to the correct answer. It is okay if you can't solve it but don't be afraid to try different methods. Don't just write the solution. Include all the things you tried that didn't work, all the stuff you normally erase or have on scrap paper that you crumple up.

(1) Three men check in at a motel, and the manager charges them $30 for a room. They split the bill and each pay $10. Later the manager realizes that the cost should have been $25. The manager gives the bellboy five $1 bills and tells him to return it to the men. The bellboy realizes that he cannot evenly divide the $5 so he keeps $2 and gives each of the men one dollar back. Now each man has paid $9 for the room, giving a total of $27. The bellboy kept $2, for a total of $29. Where is the missing $1?

(2) Three men wearing hats are standing in a row. Each person can only see the hat color of anyone in front of them. They know that their hats come from a set of two red hats and three blue hats. Each person has to try to figure out what color hat he has on. The last person says he does not know. Then the middle person says he does not know. The first person knows and calls out the correct color. What color hat is he wearing and how did he know?

(3) Two women acquaintances meet at a store. "If I remember correctly you have three sons," says Leah. "How old are they?" "The product of their ages is 36" says Beth "and the sum of their ages is exactly today's date." "I'm sorry, Beth" Leah says after thinking "but that doesn't tell me the ages of your boys." "Oh, I forgot to tell you that my youngest son has blue eyes." Leah says, "I now know exactly how old your 3 sons are." How did Leah figure out their ages?

Chapter 6

Study Techniques

6.1 Types of learning

All of the study tips in this chapter come from experts in the field of learning how to learn. If you would like to read about these topics in more depth, any of the following books are great reads: *Make it Stick* by Brown, Roediger, and McDaniel; *A Mind for Numbers* by Barbara Oakley; and *How We Learn* by Benedict Carey.

First, a little background on how our brains learn. There are two types of thinking: focused thinking and diffuse mode. Focused thinking is when you are concentrating and actively working. This occurs when you are practicing problems that you already know how to do. For example, you have already learned algebra so when solving an algebra problem you would be in focused thinking mode as you work out the solution. Diffuse mode is more passive. You let your mind wander and relax your attention. You aren't actively trying to solve a problem but your mind is subconsciously still working on it. You need to use this mode when learning a new topic. To get yourself into this mode you can take a break and let your mind relax. Both of these states of thinking can't occur at the same time, but they are both crucial for learning new concepts. The following exercise is commonly used to understand the difference between focused and diffuse thinking.

Exercise 6.1 Find the three errors in the following sentence: Thiss sentence contains threee errors.

_____

_____

_____

Finding the first two errors required focused mode, but finding the third required diffuse mode. The takeaway from this information is that it is important to allow yourself time to work on problems and don't think that you can do all of your homework in one shot the night before it is due.

Many people have a false belief that reading material multiple times is a good way to study. This has proven to be false. In one study researchers found that there was no added benefit to re-reading material and the only time where there was a slight benefit was when the additional reading took place after a significant lapse of time [Callender and McDaniel, 2009]. There was no definition for how long a "meaningful" time lapse was, but it definitely included a period of sleep. This further solidifies that allowing multiple days to work on homework and study for tests is ideal. This idea of a time lapse is called spaced repetition. It allows items in your brain to move from short term memory into long term memory.

Why do so many people think that re-reading material is a good study technique? It gives you an illusion of mastery. You become familiar with the reading and this makes you falsely think that you also understand the material. Have you ever sat through a class, nodding along, feeling like you completely understood what the teacher was saying, only to find that when it came time to do the homework you had no clue how to do it? That's because you had an illusion of mastery. Other common study techniques that produce this are highlighting and underlining [Dunlosky, Rawson, Marsh, Nathan, and Willingham, 2013]. People don't learn by being passive consumers; they learn by actively participating. You can try using index cards to quiz yourself on the material. Online index cards, such as quizlet, can be very useful tools. You can read a paragraph then try to paraphrase it into your own words. After a class you can try re-writing the class notes without looking back at them. You can try to relate new material to old material that you already know. As you study you can talk out loud to yourself.

Exercise 6.2 What are some other ways you can think of that allow your learning to be active rather than passive?

_____

_____

_____

_____

_____

_____

We have seen that it is important to take breaks as a way to switch from focused to diffuse thinking to help your brain better learn material. Sleep is an important type of break, but there are other ways to take a break that could be beneficial. Any type of exercise is a great idea. Many great writers, such as Charles Dickens, Roald Dahl, and Virginia Woolf, were also avid walkers. They knew that going for a walk is a great way to switch to diffuse mode. Similarly, any type of mindfulness practice such as meditation or yoga will work well. Other break ideas include coloring, being silly, blowing bubbles, and drinking tea.

Exercise 6.3 What are some break ideas that you think will work for you? They can come from these suggestions or you can come up with your own.

_____

_____

_____

_____

_____

_____

After you learn a new topic it is extremely important that you practice it so that you strengthen the pattern in your brain and help move that concept into long term memory. That's the reason why homework is assigned. However, not all practice is equal. In fact, misplaced persistence is actually harmful. Just because you do a problem set doesn't automatically mean that you are learning. If you don't have enough understanding of the concept and end up practicing the wrong method over and over, this will cause harm because you are encoding incorrect information in your brain. People often use the phrase "practice makes perfect." A more correct version of this phrase is "practice makes permanent."

Exercise 6.4 Explain what this student is doing incorrectly. Would this student be learning anything if asked to complete a worksheet full of these types of problems for homework?

$(3x+4y)^2 = 9x^2 + 16y^2$
$(2x+y)^2 = 4x^2 + y^2$
$(2x+3y)^3 = 8x^3 + 27y^3$

_____

_____

_____

_____

_____

    In one study baseball players were split into three groups; one group did extra batting practice for six weeks. Each practice session they got 45 extra pitches that were evenly split up across fastballs, curveballs, and changeups. Another group also got 45 extra pitches each practice session but the type of pitch was random. The third group was the control group and they got no extra practice [Hall, Domingues, and Cavazos, 1994].

Exercise 6.5 Which group do you think did better after those six weeks and why?

_____

_____

On a pre-test it was shown that all three groups performed similarly. Not surprising, on the post-test both of the groups that received extra training did better than the control group that received no extra training. The group that received random pitches did better than the other group both in a test of random pitches and a test of pitches thrown in blocks of pitch type. During the six week practice time the group that got blocked pitches at first seemed to get better quicker. The randomly pitched group progressed slower, but the gains were longer lasting.

Using the context of sports, this result may not seem surprising, but now think about how classroom learning works. Typically in a math class you learn one topic, practice it during homework, and the next class move on to a new topic. This is using the blocked approach. This approach is used for a variety of reasons and it would be difficult and possibly confusing for a class to be taught in a randomized approach. There is a technique some teachers use called spiraling that is a move towards the random approach. In a spiraled classroom, students see the same topics repeatedly at different times during the course. The material in this course is somewhat naturally spiraled because we first learn definitions for basic number theory concepts and then do different types of proofs using those same concepts. You will also see that the home-work in the proof chapters is spiraled. Just like in this study, you may not like it at first because it may seem to slow your progress, but by the time the midterm or final comes up, you will appreciate that you have been studying all along and find that you already know the material. This idea of mixing the different topics together is called interleaving.

Exercise 6.6 Would you prefer to have homework be on just the topic you learned that day or would you prefer for homework to be inter-leaved? Why?

In Chapter 2 we learned a lot about group work and its many benefits. Another reason why group work is beneficial is because whenever you discuss what you learned it helps to clarify and make ideas stick in your mind so that you can remember them better. Interestingly, you don't need to have mastered the material in order to get this benefit. You probably have found that speaking out loud about problems, even without input from anyone else, can help you put your thoughts together and draw conclusions. Understanding often arises from this step of speaking your thoughts out loud and trying to explain to others. This is one of the reasons why forming study groups can be beneficial.

6.2 Homework

Exercise 6.7 What do you think the purpose of homework is?

_____

_____

_____

_____

_____

_____

It is generally agreed that some of the purposes of homework are to practice material just learned, to apply the topic just learned to a new context, to prepare for a future topic, to integrate multiple topics learned, or for a non-academic purpose such as learning organizational skills. The purpose of homework in this class will be to either reflect on a topic we went over in class or to practice and integrate topics from previous classes. John Dewey is quoted as saying "We do not learn from experience… we learn from reflecting on experience." In Chapter 1 we discussed using reflection as a way to help understand what level of understanding you are at. In addition, we see it can be important to reflect on learning to solidify ideas and make the learning stick. Homework from the earlier chapters was mainly reflection type questions. Once we get to the chapters on proofs, the homework will mainly be on practicing skills. The questions are chosen to encourage deliberate practice. This is a term coined by the psychologist Anders Ericsson to describe a type of practice that involves a purpose and a focus. This is the opposite of mindless repetition in practice. One of the main components of deliberate practice is to receive feedback and make corrections. For this reason you will not see homework pages like a typical textbook with many problems that are repetitive and encourage mindless repetition. Instead, you will find a small amount of problems that are specifically chosen to get the most bang for your buck. You are a busy person, probably taking many classes. There is no need to keep practicing a concept if you already understand it. If you don't understand a problem then the homework is a time to reflect on that, discuss with classmates, and get feedback. Maybe you won't fully get it on this assignment, but the spiraling of homework allows for many opportunities in later homework to try again.

You may wonder why the purpose of the homework is being explained. It usually isn't ever discussed and, as a result, professors and students are not always on the same page about completing homework. Some students see it as busy work and don't deem it necessary. Through the preceding explanation it is made clear that, at least for this class, that is not the case. Other students may want to complete homework but didn't understand the topic in class and so they are unable to. That is the purpose of receiving feedback so that you can learn from any mistakes or misconceptions you have. Others, for various reasons, know that home- work is counted towards their grade and choose to copy answers from a classmate or look up answers online. The preceding argument hopefully is able to convince you that the homework in this class is key to helping you understand the material and that cheating on it is really only cheating yourself. In the short term this may lead to a successful grade, but in the long term you will not do as well on exams and you will not have learned proofs well enough to be successful in later math courses.

Exercise 6.8 Read and sign the following homework pledge.

I promise that I will complete the homework on my own. I can discuss problems with classmates, but I will write up the solutions without any help. I will not knowingly let another student copy my work, nor will I copy another student's work.

Signature _____

Date _____

The main take-aways from this chapter are:

- Use active rather than passive methods to study and don't fall prey to illusions of mastery
- Spread your learning over several study sessions with periods of sleep in between
- Spiral course material to the best of your ability
- Take breaks
- Complete the homework to the best of *your* ability

Exercise 6.9 What are some specific ways that you can apply what you have learned in this chapter to course material?

_____

_____

_____

_____

_____

_____

_____

Chapter 7

Pre-proofs

7.1 Math terms

Exercise 7.1 Try your best to define each of the terms listed below.

Axiom:

Theorem:

Lemma:

Corollary:

Conjecture:

If you have taken a Geometry course then you have probably heard the term "axiom." An axiom is a starting point. Mathematics begins with a set of axioms that everyone assumes are true so that we can build on them and prove more math facts. In Euclidean Geometry there are five axioms. For example, one of them says that all right angles are equal. These axioms we assume to be true and you will never be asked to prove them. Axioms are generally agreed upon as good starting points. Although, in Geometry there is one interesting axiom called *The Parallel Postulate* (here the term postulate is used interchangeably with the term axiom) that for a long time some mathematicians believed was not an axiom, but something that was provable from the other four axioms. If this sounds interesting, you can take a course in Non-Euclidian Geometry where variations of the fifth axiom are used to create different types of geometries.

Axioms can be used to prove other statements. Those other statements that you would prove are called theorems. For example, one of the theorems given in *Euclid's Elements* says: If two triangles have two sides equal to two sides respectively, and if the angles contained by those sides are also equal, then the triangles will be equal in all respects. This theorem had to be proved using the axioms, definitions, and previously proven theorems. Sometimes when you are trying to prove a theorem you realize that it would be helpful to have a result that you haven't yet proven. Rather than writing a proof within a proof, it makes more sense to go to the side and write out this smaller proof. We call this a lemma.

Figure 7.1. Lemma

Sometimes when you prove a theorem you can get an additional result without doing too much extra work. This is called a corollary.

Figure 7.2. Corollary

You may remember learning (and possibly also proving) the Law of Cosines: If there is a triangle with sides a, b, and c and Angles A, B, and C then

$$a^2 = b^2 + c^2 - 2bc\,(\cos A). \qquad (7.1)$$

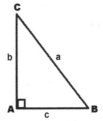

Figure 7.3. Right triangle

This is a theorem that can be proven. Once we have proved the Law of Cosines is true then we can easily prove the Pythagorean Theorem. If A is the ninety degree angle then $\cos A = 0$. The formula becomes

$$a^2 = b^2 + c^2. \tag{7.2}$$

That is an example of a corollary because it was easy to prove once we did the hard work of proving the Law of Cosines.

A conjecture is a statement that may or may not be true. We don't know because it hasn't yet been proven, but it also hasn't yet been disproven.

Exercise 7.2 Pick any positive integer. If it is even then divide it by two. If it is odd then multiply it by three and add one. Take the new number you got and repeat this step. Keep going like this until you eventually get the number 1.

This is called the Collatz conjecture. It was first posed in 1937 by Lothar Collatz. It is still called a conjecture because it has neither been proven nor disproven. We still don't know if it works for every positive integer. If this idea of unproven statements intrigues you then you may be interested in looking at other famous conjectures. One of the most famous is Goldbach's conjecture, which states that every even integer larger than 2 can be expressed as the sum of two prime numbers.

Sometimes a conjecture can turn into a theorem. In 1637 Fermat wrote a conjecture in the margin of one of his books. It said that the equation

$$a^n + b^n = c^n \tag{7.3}$$

has no integer solutions when n is bigger than 2. For a long time this remained a conjecture until 1994 when it was proven to be true by Andrew Wiles, a professor at Princeton University. An interesting aside to this story is that many mathematicians throughout history have worked on this problem. In 1847 two French mathematicians, Cauchy and Lamé, each thought they had proved the theorem using complex numbers. Their proof relied on all complex numbers being able to be factored into a unique product of primes. While this is true for real numbers, a German mathematician named Ernst Kummer showed that it does not always hold for complex numbers. As a result Kummer created a mathematical object called ideals, that helped lead to a subject called abstract algebra. Even though the French mathematicians were not able to successfully prove the theorem they were intending to, their mistake led to a whole new branch of mathematics!

One way to get a feel for how mathematicians think is to play a game discussed by Dan Finkel. He calls it *Making and Breaking Conjectures*. One person makes a conjecture and another person tries to come up with a counterexample to show that the conjecture is false. This leads to the first person refining the conjecture. This back and forth can occur as many times as needed. Here is a mock example of what this game looks like. We will start with a non-math example to make it more relatable.

Conjecture 1: Bears have 4 legs and zebras have 4 legs. I think that all animals have 4 legs.

Response: What about birds? Birds have 2 legs.

Conjecture 2: My new conjecture is that all animals have 2 or 4 legs.

Response: That is not true. Snakes have no legs.

Conjecture 3: My new conjecture is that all animals have at most 4 legs.

Response: What about centipedes?

Conjecture 4: My new conjecture is that all animals have an even number of legs.

This process is similar to the scientific method in that you make a conjecture (similar to what scientists call a hypothesis), test to see if it is accurate, and refine as needed.

Exercise 7.3 Come up with a counterexample that disproves the conjecture that multiplying a number by 10 inserts a zero at the end of the number.

7.2 Proof tips

Now we will begin to think about how to prove a theorem that is given to us. In proving theorems you use the same kind of method as you do for showing that a logical argument is valid. We will begin with the topic of number theory because you have had many years of practice working with numbers and have a general understanding of how they behave. To begin with, many of the theorems we will prove will be statements that you have used the result of before. You probably have just never proved them. To prove a theorem you can use any commonly known properties of numbers, such as that integers are closed under addition (when you add two integers together the result is also an integer), as well as any previously proved theorems and any definitions. You will begin to see that a key first step in a mathematical proof is to understand what you are given and what you are being asked to prove and to be able to correctly apply any needed definitions.

This chapter is all about mentally preparing you to write a mathematical proof, but we won't actually get to the proof writing until the next chapter. You might wonder why we need all this build up. Even though you are now taking an introduction to proofs course, this won't be your very first encounter with mathematical proofs. At the very least you should have seen some in high school geometry. If that was your only experience with proofs, you may be in for a surprise. Typically geometry proofs are taught as two-column proofs. That means there is a column for a statement and a column for the reason. This is one type of structure that in some ways is helpful, such as it reminds you that every new statement made must have some reason attached to it. But in other ways this structure is harmful because it decreases creativity and reinforces the false belief that there is only one correct solution.

In Chapter 1 we discussed how most people view mathematics as a subject of formulas and procedures whereas mathematicians view mathematics as creative pattern detection. If you fall into the first category then it is a big shift to begin changing your view.

In *A Mathematician's Lament* Paul Lockhart satires the current state of math education by showing how music would look if it was taught like math. Students would initially start by learning all of the rules of sheet music and music theory. Students wouldn't actually be allowed to hear or play music until graduating high school, and only music majors would really get to. Lockhart uses this analogy to exaggerate the fact that many students don't get to experience the creative joy of mathematics because they are too bogged down by having to memorize and practice all of the mechanics of mathematics. This introduction to proofs course is often seen as a gateway to this new way of thinking about math that focuses on creative problem solving.

This course is not your first encounter with proofs. We already discussed how they were used in a geometry class. You may have also seen some proofs in a calculus course. Calculus is generally taught as an applications-based course. Calculus proofs are taught in a course called real analysis and we will get a sampling of them in the last chapter of this book. In the calculus course you took, the professor most likely wrote out some proofs rather than asked you to prove them. Or maybe in high school you saw a proof of the Pythagorean Theorem. There are many different ways to prove this theorem; in fact, there are whole books written about this. But a typical proof looks like this:

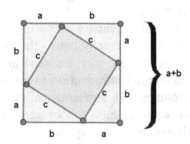

Figure 7.4. Pythagorean Theorem proof

Each side of the larger square has a length of $a + b$ because it contains one side of length b from one triangle and one side of length a from another triangle. This makes the area of the larger square $(a + b)(a + b)$. The area of the inner square is c^2. The area of the four outer triangles is equal to $4(1/2\ ab)$, or $2ab$. This means that

$$(a + b)(a + b) = c^2 + 2ab. \tag{7.4}$$

When you multiply out that equation and rearrange its new terms you end up with

$$a^2 + b^2 = c^2. \tag{7.5}$$

Exercise 7.4 What are your initial thoughts on this proof?

_____

_____

_____

_____

Most likely you were able to follow along with each step that was presented. All of the steps were fairly simple. All we used were the formulas for the area of a square and a triangle, as well as some algebra. So it's not that the proof itself is difficult to understand. What baffles most students and makes proofs seem out of reach is the starting point. Where did that image come from? Why were squares involved in the proof of a theorem that only involves triangles? These are valid questions that aren't explained by the proof itself. To add to the mystique, when the proof was presented to us we only saw the final, polished version. It is highly unlikely that the first person to think of this proof immediately thought to use this picture and everything worked out perfectly. Here is an example of what might have occurred the first time someone tried to prove the Pythagorean Theorem.

We want to prove $a^2 + b^2 = c^2$.

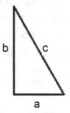

Figure 7.5. Triangle

The solver knew the formula has squared terms so they reasoned that it might make sense to draw a square using two copies of the triangle.

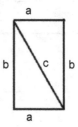

Figure 7.6. Rectangle from two triangles

This is a rectangle unless $a = b$. To simplify things let's look at the case where $a = b$ and see if we can generalize from there.

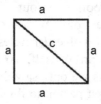

Figure 7.7. Square from two triangles

When $a = b$ the rectangle is a square with an area of a^2 and the area of each triangle is $\frac{1}{2} a^2$. That would make the area of the two triangles

$$\frac{1}{2} a^2 + \frac{1}{2} a^2 = a^2. \tag{7.6}$$

We can observe that was the same as the area of the square but it didn't lead us anywhere. Let's try something different. We need to get a "b" and "c" term in there somehow. Maybe we can make a square out of four of the triangles.

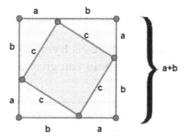

Figure 7.8. Square with four triangles

This will work because the length of each side is $a + b$, making it a square. And the proof presented earlier will now go here.

In class professors usually only show the final, perfect proof with no mistakes. This occurs for various reasons, such as it would waste time to show the complete thought process, or it might confuse students if they see too many mistakes. The professor has probably done this proof so many times that it has been memorized and can easily be reproduced. Unfortunately, though, the professor probably never explained that the very first time he or she was asked to prove this theorem it involved struggle and mistakes. You won't know that the professor had to persevere through repeated failures and eventually, possibly with help from a colleague or a mentor, was able to complete the proof. I strongly believe that this is the reason so many students think that they are bad at proofs or they "just don't get it." The secret is that nobody "just gets it." Just because you can't look at a proof and immediately have everything fall into place does not mean that you are bad at proofs. You can't be afraid to explore a problem. Sometimes you will choose the right path, but other times your path leads nowhere. If you have embraced the growth mindset ideas we discussed earlier then you know that doesn't mean you should give up; it simply means that you need to try a new path.

You can use some of the study tips we discussed, such as if you are having difficulty with a proof it is a good idea to take a break and try working on it again another day. Neuroscientists have demonstrated that our brain subconsciously works on tasks even when we aren't consciously thinking about them. By taking breaks and letting our subconscious take over we conserve attention and willpower that would be exhausted if we tried to struggle past our limits. Everything we have learned up until this point was to prepare us to be able to write a mathematical proof. You now have many tools in your tool belt that you can utilize when working on a difficult problem.

Exercise 7.5 One of your classmates is trying to prove something but he is having trouble. What advice would you give him?

_____

_____

_____

_____

_____

_____

A student who previously took this course was asked to share her classroom experience with writing proofs. Here is what she wrote.

In this course you are going to learn a variety of mathematical proofs. In the beginning of this course you will start off with motivating powerpoints and videos. This will prepare you for what is to come later on in the semester. The most important thing to do is NOT give up. Some proofs may look very challenging but when you start off with the foundation of how to do a proof you will apply that to multiple proofs and WILL be able to do it. A mathematical proof is essentially proving something that you may or may not know to be true. You will apply what you know and what the givens are to conclude a statement that is true. In the beginning you will start off with simple proofs that don't require many steps. You will use those techniques and foundation on multiple other proofs and expand. You will learn how to recognize what the problem is asking, how to write it in simpler terms and then plug in what you know and work with it to get what you are trying to prove. You will be working in groups multiple times and it helps because you get to hear others thoughts and ideas and work together to complete the problem. This was very beneficial because on some problems I didn't even know how to start them but working with my group gave me that start so I can go on and finish the problem. The homeworks in this class should be a HUGE grade booster. I loved having the opportunity to go back and forth with the professor until I understood the problem. With the support of the professor and classmates, you will be able to solve mathematical proofs.

Figure 7.9. Note from former student

Exercise 7.6 After reading this note, how do you feel about the next phase of this course where we begin proofs?

One other helpful thing to keep in mind before we get to the proofs chapters is that unlike many computationally-based classes, your professor may not immediately know the solution. That's not to say that your professor is not capable of finding a solution. Rather, he or she may be unable to look at a problem and instantly see what strategy to use, but with ample time and energy would be able to solve it. Believe it or not, this is actually a good thing. As we just learned, nobody is born knowing how to write a perfect proof with everything falling into place every single time. If you think your professor can effortlessly solve all proofs then you are perpetuating, possibly even unintentionally, the false belief that some people "just get" proofs.

Another downfall to your professor knowing how every proof works ahead of time is that it makes it tempting for him or her to give away too much. This can take away the experience that you need to go through as you struggle and make mistakes. As we saw with the Pythagorean Theorem proof, just because you understand a proof doesn't mean that you know how to write a proof. It is easy to fall into the typical paradigm where the teacher's job is to give information and the student's job is to receive information. That may work well in other math classes but it will not work in an introduction to proofs class. The professor can model how to prove every single theorem given in this book and you can still walk away not knowing how to create your own proof. You will need the chance to explore and make mistakes and you need to figure out the path on your own.

Chapter 8

Direct Proofs (Even, Odd, & Divides)

8.1 Definitions

After all that buildup we are finally ready to start proving theorems. One of the key first steps to writing a proof is to make sure that you fully understand all of the terms used in the statement you are proving. If you don't understand what the statement is asking you to prove then it will be extremely difficult to proceed. To that end, we will start off by going over some definitions. Many of these terms will be familiar to you, but you may have never developed an explicit definition for them. This step also ensures that we are all on the same page, as sometimes a term can be defined in more than one way.

Exercise 8.1 Give some examples of even numbers. How would you define an even number?

_____

_____

_____

_____

The formal definition that we will use for an even number is as follows:

Definition 8.1 n is even if and only if there exists an integer k such that $n = 2k$.

To show that a number is even we will have to match it to this definition. We know that 6 is an even number. In order to show this we would need to re-write 6 as $2(3) = 2(\text{integer})$. This exactly matches the definition for being even.

Exercise 8.2 If x is a given integer then show that $4x + 4$ is even. Show that 0 is even.

Exercise 8.3 Give some examples of odd numbers. How are odd numbers related to even numbers?

_____

_____

_____

_____

The formal definition that we will use for an odd number is as follows:

Definition 8.2 n is odd if and only if there exists an integer k such that $n = 2k + 1$.

Exercise 8.4 Use the definition to show that 7 is odd and $6x + 3$ is odd given that x is an integer.

For integers a and b we will read $a|b$ as "a divides b." The formal definition that we will use for an integer dividing another integer is as follows:

Definition 8.3 $a|b$ if and only if there exists an integer k such that $b = ak$.

Exercise 8.5 Use this definition to show that $3|6$ and $2|4m$, where m is an integer.

Exercise 8.6 Given a and b are integers, how are the following expressions the same and how are they different?

$$a|b \qquad\qquad \frac{a}{b}$$

_____

_____

_____

_____

_____

_____

8.2 Proofs

Let's get started with our first proof. The process we will use works the same way as when we completed the exercises showing that a logical argument is valid. We used our givens along with the known logical arguments (which we can think of as axioms) to make a logical conclusion. Each step followed from the one before it, forming a chain of reasoning that eventually led to what we wanted to conclude. If you get bogged down in the details of a proof, try to keep this big picture in mind.

Example 8.1 Prove: Any even number times any odd number is even.

The first thing we will do is re-write the statement using a quantifier and write it as a conditional statement. This step is not always necessary;

sometimes the proof is already given in this form and sometimes it doesn't fit the if-then format. The majority of the statements we want to prove will be "for all" statements. In this example we know it is this type of statement because of the word "any," but even if there is no quantifier term, the universal quantifier will be implied. In this statement we need to define two numbers, one even and one odd. Then we want to prove that their product is even. Here is how it will look in the correct format.

Prove: $\forall\, x, y \in \mathbb{Z}$, if x is even and y is odd then xy is even.

I chose to represent my even and odd numbers as x and y, but any variable you use is fine. It also didn't matter which was called x and which was called y. With the quantifier I chose to put x and y in the set of integers. The reason for this is that we know they are even and odd numbers. Think about what set even and odd numbers come from and it will make the most sense to use the set of integers. It wouldn't have been wrong to say the set of rational numbers or the set of real numbers, but using integers will be more precise.

For the next step let's think about what we know. The nice thing about putting your statement into the conditional form first is that it is very clear what your givens are. Anything that is between the "if" and the "then" will be given. In this example I know that x is even and y is odd. What we want to do is get the givens into a useable form. In most cases, that means applying definitions. We recently defined an even number as 2 times an integer and an odd number as 2 times an integer plus 1.

Exercise 8.7 Why is it incorrect in this example to say $x = 2k$ and $y = 2k + 1$ for some integer k?

_____

_____

_____

_____

To put the givens in useable form I will say $x = 2k$ and $y = 2j + 1$ for some integers j and k. Any time you introduce a new letter you will need to specify what it stands for. In this chapter we are only dealing with integers, but later on we will also look at real numbers. Even if you think it is obvious what your letter should stand for, try to keep in mind that when you write a proof it is a form of communication and that can only be achieved if everyone understands your notation. As you progress into higher level math classes your professor may relax this condition. For example, in a number theory course we typically deal with integers so common practice is to say anytime you introduce a letter we will assume it stands for an integer unless otherwise noted. But for now, especially since you are just starting out, let's aim to be clear.

For the next step, think about what you want to prove. Again, the beauty of the conditional statement is that what you want to prove is what follows the "then." In our example we want to prove that xy is even. But again, think about this in a more usable form. What does it mean to be even? We need xy to be 2(integer). I would suggest not using a letter here because it will come from our givens and we don't know what form that integer will take until we do the proof out.

For the next step, think of a plan of action. As we get further into the course we will learn many different proof methods and this is a good point to think about which method will work. As we discussed in Chapter 1, it's okay to make mistakes. Perhaps you think one method will work and it turns out not to. That is not a big deal; it just means you will come back to this step and try something different. For the next four chapters we will just be discussing direct proofs. That means that you will directly prove the statement from your givens. After that we will learn other proof techniques.

Now, you will play around with your givens and see what happens. I call this the doodle stage because it is similar to when you doodle a picture. Merriam Webster Dictionary defines a doodle as "an aimless or casual scribble, design, or sketch." Think of this stage as exactly that. You are free to try out ideas and see where they lead you. Maybe you will get what you want to prove, but maybe not. If not, go back and try a different idea. This is the productive struggle stage. This is the stage that is obscured when a polished proof is presented to you. Some proofs might just naturally fall into place, but if they don't you need to give yourself permission to work through that struggle. There's no single way to go through this step, but one suggestion is to try out some examples and convince yourself that the statement is really true. Our problem starts with an even and an odd number. As an example let's choose 2 and 3. Then we need to multiply those. The product will be 6. We know 6 is even because it can be written as 2(3), which is 2(integer).

Exercise 8.8 We showed that the statement was true when $x = 2$ and $y = 3$. Why is that not considered a valid proof for this statement?

_____

_____

This example worked out, but if you're still not convinced you can try out some more examples. Oftentimes working through some examples also gives you a good idea of how the proof will work. My plan will be to multiply out x and y and try to factor out a 2 like we did in the example. This proof will be fairly straightforward, so there is not much else to do in this stage.

Let's move on to the final stage, which is writing the final proof. We will follow the format for the example but make it for the general case. Start with our givens, $x = 2k$ and $y = 2j + 1$ where k and j are integers. Now multiply to get $xy = (2k)(2j + 1)$. We want to show this is an even number, just like in our example where the product was 6. There are multiple ways to do this. Some people may want to multiply the values to get $4kj + 2k$. Then we can factor out a 2 to get $2(2kj + k)$. Since k and j are integers and we know integers are closed under addition and multiplication (i.e. if you add or multiply any two integers the answer is still an integer), we also know that $2kj + k$ is an integer. Thus, we have proven that $xy = 2(\text{integer})$. This is our definition for being even. Somebody else may have noticed that $(2k)(2j + 1)$ can be thought of as three numbers multiplying, $2[(k)(2j + 1)]$. If you think of $k(2j + 1)$ as the integer, then we have gotten this into the correct form for an even number. Typically when a proof is completed the author will denote this by putting a small square, a checkmark, or the letters QED (an abbreviation for the Latin phrase meaning "what was to be shown") at the end. It seems like a small thing, but adding one of those when you have completed a proof can be a big source of satisfaction. Here is what my final proof looks like:

Prove: $\forall\, x, y \in \mathbb{Z}$, if x is even and y is odd then xy is even.

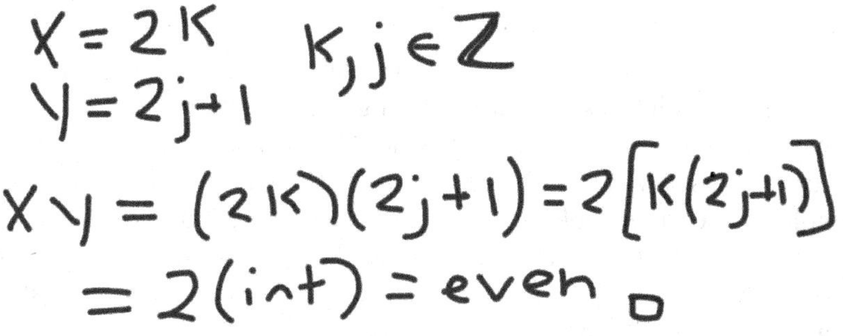

$$X = 2K \qquad K, j \in \mathbb{Z}$$
$$Y = 2j+1$$
$$XY = (2K)(2j+1) = 2\left[K(2j+1)\right]$$
$$= 2(int) = even \quad \square$$

Figure 8.1. Proof that product of even and odd is even

Example 8.2 Prove or disprove: The sum of any odd number and 3 times any even number will be even.

First we will get this into the formal notation.
$\forall\, x, y \in \mathbb{Z}$, if x is odd and y is even then $3y + x$ is even.
Next, we will put our givens into useable form. We know that x is odd so we will write it as $x = 2k + 1$, where k is an integer. We know that y is even so we will write it as $y = 2j$, where j is an integer. At this point we do not know if the conclusion that "$3y + x$ is even" is true or false. The plan will be to start with some examples to see if it looks to be true or false. Choose $y = 2$ and $x = 3$. Then $3y + x = 3(2) + 3 = 6 + 3 = 9$.

Exercise 8.9 Why does letting $y = 2$ and $x = 3$ to get $3y + x = 9$ prove that the statement is false?

_____

_____

_____

To disprove a statement you just need to find one counterexample; that is, an example that shows that the statement is false. If a statement is false for just a single case then that means it can't be true for all cases.

Example 8.3 Prove or disprove: For all integers a and b, if a|b and a|c then a| (5b − 2c).

This statement is already given in the formal notation so the next step is to put the givens into useable form. Using the definition for divides they can be written as b = ak and c = aj, where k and j are integers. We want to prove or disprove that 5b − 2c = a(integer). The plan is to try some examples to see if it looks true or false. If it looks true we will try to prove it using the definition of divides. We know a|b and a|c so the first example I will choose a = 2, b = 4, and c = 8. These values will satisfy the givens (2 divides both 4 and 8). Next, look at 5b − 2c. This will become 5(4) − 2(8) = 20 − 16 = 4. It is true that 2 divides 4 so it looks like the statement could be true. This is not a proof of it, though, because there may be other values where the givens are both true but the conclusion is not true. At this point you can choose to try out more examples or you can move on to trying to prove the statement. Even if you thought the statement was true and it turned out to be false, as you try to do the proof you will come to a point where nothing seems to work and that will be your clue that it may actually be false. This example does happen to be true so we will continue on to the proof.

Proof.

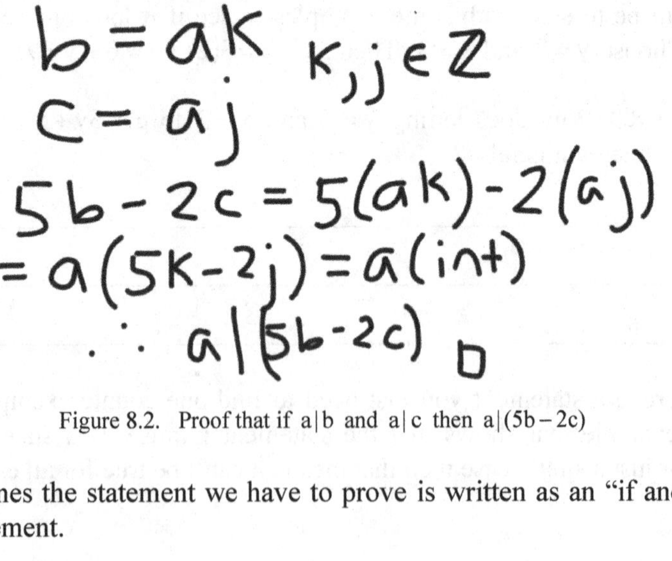

$$b = ak \quad k, j \in \mathbb{Z}$$
$$c = aj$$
$$5b - 2c = 5(ak) - 2(aj)$$
$$= a(5k - 2j) = a(int)$$
$$\therefore a \mid (5b - 2c) \quad \square$$

Figure 8.2. Proof that if a|b and a|c then a|(5b − 2c)

Sometimes the statement we have to prove is written as an "if and only if" statement.

Exercise 8.10 If you were asked to prove: $\forall\, x, y \in \mathbb{Z}$, x is even if and only if x^2 is even, how can you break this up into the two parts that are needed to be proven?

Exercise 8.11 Look over the following four "proofs" and decide whether each one is valid or invalid. Once you feel confident in your decision, consult with a partner and convince him or her why you are correct. All four examples are for the following:

Prove or disprove: $\forall\, a, b, c \in \mathbb{Z}$, if $a|bc$ then $a|b$ and $a|c$.

(1) Proof: $bc = ak$, where k is an integer $\rightarrow b = \frac{ak}{c} = a(\text{integer}) \rightarrow a|b$. Also, $c = \frac{ak}{b} = a(\text{integer}) \rightarrow a|c$. $\square$

(2) Proof: Let $a = 2$, $b = 2$, and $c = 4$. It is true that $2|8$ and it is also true that $2|2$ and $2|4$. $\square$

(3) Disproof: Let $a = 2$, $b = 3$, and $c = 5$. 2 does not divide 3 and 2 does not divide 5. $\square$

(4) Disproof: Let $a = 2$, $b = 3$, and $c = 4$. It is true that $2|12$ but 2 does not divide 3. $\square$

Exercise 8.12 Do the following exercises. The framework is provided for each proof.

(1)   Statement to prove or disprove:

Any even number plus any odd number is odd.

If needed put in ∀ _____ if _____ then _____ form:

What do you know? (if needed put in useable form):

What do you want to prove? (put in useable form):

What is your plan?:

Doodle (play around with your givens and try to get what you want to prove):

Final Proof:

(2)   Statement to prove or disprove:

$\forall\, a,b \in \mathbb{Z},$ **if a is even and b is odd then** $\frac{a^2+b^2+1}{2}$ **is an integer.**

What do you know? (if needed put in useable form):

What do you want to prove? (put in useable form):

What is your plan?:

Doodle (play around with your givens and try to get what you want to prove):

Final Proof:

(3)   Statement to prove or disprove:
Every integer divides itself.

If needed put in ∀____ if _____ then _____ form:

What do you know? (if needed put in useable form):

What do you want to prove? (put in useable form):

What is your plan?:

Doodle (play around with your givens and try to get what you want to prove):

Final Proof:

(4) Disprove: $\forall$ **a,b** $\in \mathbb{Z}$, **if a|b and b|a then a = b.**
Then change it so the statement is true and prove it.

If needed put in $\forall$ ____ if _____ then _____ form:

What do you know? (if needed put in useable form):

What do you want to prove? (put in useable form):

What is your plan?:

Doodle (play around with your givens and try to get what you want to prove):

Final Proof:

Exercise 8.13 Write a reflection on what you learned from any of the mistakes you or a classmate made on any of the proofs you did today.

_____

_____

_____

_____

_____

_____

Chapter 8 Homework

Use the framework or a similar process to solve each problem. A blank copy of the framework can be found at the end of the book in Appendix G. You may use this to make photocopies. In future chapters the framework will not be listed in the work area like it was in this chapter, but it is suggested that you continue to utilize it until you feel comfortable with the process. For each homework keep in mind all of the study tips that we went over in Chapter 6. One key idea is that you need to take breaks that include a period of sleep. To that end, you will find a homework log in the back of this book in Appendix H. Use that page to keep track of what days and approximately how long you spend working on each homework. Another key idea is to interleave topics. As we learn more definitions you will see that the homework is interleaved in the sense that proofs will come from any of the previously learned definitions.

(1) Prove or disprove: Any even number times any even number is even.
(2) Prove or disprove: The product of any two even integers is even if and only if both are even.
(3) Prove for all integers a, b, x, and y, if $a|b$ then $a|(ax+by)$.
(4) Prove or disprove: 5 divides the sum of any 5 consecutive integers.
(5) Explain the process of how to write a mathematical proof in your own words.
(6) Explain the difference between "divides" and "divided by." Use examples to help explain.

Chapter 9

Direct Proofs (Rational, Prime, & Composite)

9.1 Definitions

Exercise 9.1 Give some examples of rational numbers. How would you define a rational number?

_____

_____

_____

_____

The formal definition that we will use for a rational number is as follows:

Definition 9.1 r is rational if and only if there exists integers k and non-zero j such that $r = \frac{j}{k}$.

Exercise 9.2 Use this definition to show that 3 is rational. Explain what has to be true in order for $\frac{3x+4}{5y}$ to be rational.

Exercise 9.3 Explain why each of these numbers would not belong with the others.

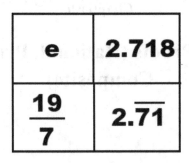

Figure 9.1.   Which one doesn't belong part II

Exercise 9.4 Give some examples of prime and composite numbers. How would you define these terms?

_____

_____

_____

_____

_____

_____

The formal definition that we will use for a prime number is as follows:

Definition 9.2 An integer $p > 1$ is prime if and only if p has no integer divisors other than 1 and itself.

The formal definition that we will use for a composite number is as follows:

Definition 9.3 An integer $c > 1$ is composite if and only if there exists integers n and m such that $c = nm$ where n or $m \neq 1$.

9.2 Proofs

Example 9.1 Prove or disprove that the sum of any two rational numbers is rational.

First, let's get this into the formal notation. $\forall\, x, y \in \mathbb{R}$, if x and y are rational then $x + y$ is rational. We know that x and y are rational so they can be written as $x = \frac{a}{b}$ and $y = \frac{c}{d}$ with a,b,c,d integers, $b \neq 0$, and $d \neq 0$. We want to prove that $x + y$ can also be expressed as an integer divided by an integer with the denominator not zero. To see whether this statement is true we can try some examples. Let $x = \frac{1}{2}$ and $y = \frac{2}{3}$. Then $x + y = \frac{1}{2} + \frac{2}{3} = \frac{3}{6} + \frac{4}{6} = \frac{7}{6} =$ rational. It looks to be true. The plan to prove it is to find a common denominator to add the rational numbers together.

Proof.

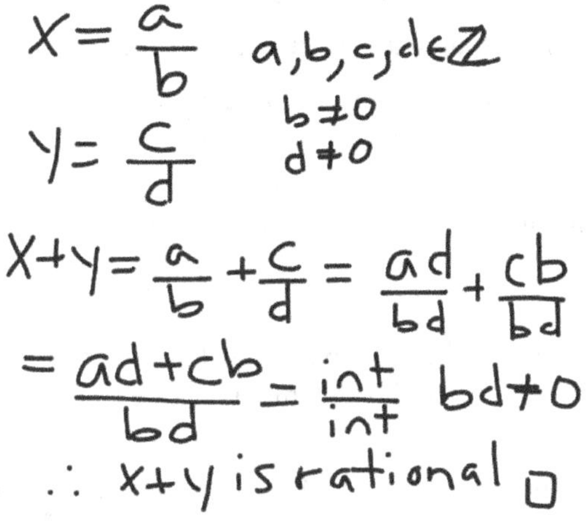

$$X = \frac{a}{b} \qquad a,b,c,d \in \mathbb{Z}$$
$$b \neq 0$$
$$Y = \frac{c}{d} \qquad d \neq 0$$

$$X + Y = \frac{a}{b} + \frac{c}{d} = \frac{ad}{bd} + \frac{cb}{bd}$$
$$= \frac{ad + cb}{bd} = \frac{int}{int} \qquad bd \neq 0$$
$$\therefore\ X + Y \text{ is rational} \qquad \square$$

Figure 9.2. Proof sum of two rationals is rational

Note: Since the integers are closed under addition and multiplication we know that $ad + bc$ is an integer and bd is an integer. We also know that b and d separately are not equal to zero, which means if we multiply them together the answer will also be non-zero. Thus, $bd \neq 0$.

Exercise 9.5 In Example 9.1, why did we have to show that $bd \neq 0$?

_____

_____

_____

Example 9.2 Prove or disprove: For all real numbers k, if k is rational then $\frac{k^3}{k+1}$ is rational.

Proof. We know that $k = \frac{a}{b}$ with a,b integers and $b \neq 0$. We are trying to prove that $\frac{k^3}{k+1}$ can be expressed as an integer divided by an integer with the denominator not zero. Thinking about how rational numbers work, it doesn't seem like raising to the third power or adding 1 will take us out of the set of rational numbers. The main concern is making sure that the denominator is not zero. The only given is that k is rational, which does include the possibility that k is -1. In that case, $k + 1 = 0$ and $\frac{k^3}{k+1}$ will be undefined. That will serve as a counterexample disproving our given statement. $\square$

Example 9.3 Change the statement in Example 9.2 so that it is true and then prove it.

We saw that the only problem arose when $k = -1$. If we add the restriction that k cannot be -1 then the statement will be true.

Prove: For all real numbers $k \neq -1$, if k is rational then $\frac{k^3}{k+1}$ is rational.

Proof. Let $k = \frac{a}{b}$ with a,b integers and $b \neq 0$. Then

$$\frac{k^3}{k+1} = \frac{\left(\frac{a}{b}\right)^3}{\frac{a}{b}+1} = \frac{\frac{a^3}{b^3}}{\frac{a+b}{b}} = \frac{a^3}{b^3} \cdot \frac{b}{a+b} = \frac{a^3}{b^2(a+b)} = \frac{a^3}{ab^2+b^3}. \tag{9.1}$$

We know that a^3 and $ab^2 + b^3$ are integers since a and b are integers and integers are closed under multiplication and addition. We also need to show that the denominator is not zero. If it were, then

$$b^2(a+b) = 0 \rightarrow b = 0 \; or \; a+b = 0. \qquad (9.2)$$

We know that b cannot be zero because it was given. If $a + b = 0$ that implies that $a = -b$, which can be written as $\frac{a}{b} = -1$. We were given that k, which we expressed as $\frac{a}{b}$, cannot be -1. This means that the denominator cannot be zero and we have proved that $\frac{k^3}{k+1}$ is a rational number. $\square$

Example 9.4 For all integers n, what are the possible values for n if $n^2 - 2n - 3$ is prime?

Thinking about the definition of a prime number, you can see that the key to most prime proofs will be the idea of factoring. The plan is to try to factor this expression and then see when one of the factors is equal to 1. Factoring gives $n^2 - 2n - 3 = (n+1)(n-3)$. To be prime either $n+1 = 1$ or $n - 3 = 1$, which means that either $n = 0$ or $n = 4$.

Exercise 9.6 Can we stop the proof at this point? If yes, explain why. If no, keep going to complete the proof.

Exercise 9.7 Complete the following. Use a copy of the proof framework or list your steps following the same format.

(1) Prove or disprove: The product of any two rational numbers is rational.
(2) Prove or disprove: The square of any rational number is rational.
(3) Find all values where $n^2 + 8n + 15$ is prime, where n is an integer.
(4) Prove or disprove: $\forall\, a, n \in \mathbb{Z}$, if $a^n - 1$ is prime then $a = 2$.

Exercise 9.8 Write a reflection on what you learned from any of the mistakes you or a classmate made on any of the proofs you did today.

_____

_____

_____

_____

_____

_____

_____

_____

Chapter 9 Homework

Fill in your homework log as you work on the following problems.

(1) Prove or disprove: The sum of any two odd integers is even.
(2) Prove or disprove: The product of any rational number and any irrational number is rational.
(3) Fill in the blank and then prove: Any rational number plus any even number is _____.
(4) Prove or disprove: $\forall a, b, c \in \mathbb{Z}$, if $a \mid b$ and $b \mid c$ then $a \mid c$.
(5) Prove or disprove: If n is a positive power of 2 then $n + 8$ is composite.
(6) Sam was asked to find all prime numbers of the form $x^2 - x - 12$. He showed the following work. Explain what is good about it, what is incorrect, and how it can be fixed to be correct. Include the correct proof in your explanation.

$$x^2 - x - 12$$
$$(x + 3)(x - 4)$$
$$x + 3 = 0 \qquad x - 4 = 0$$
$$x = -3 \qquad x = 4$$

Figure 9.3.   Homework 9.6

Chapter 10

Direct Proofs (Square Numbers & Absolute Value)

10.1 Definitions

Exercise 10.1 List the first 5 square numbers (sometimes known as perfect squares). Draw a picture to represent each one and explain why they are known as square numbers. How would you define a square number?

_____

_____

_____

_____

The formal definition that we will use for a square number is as follows:

Definition 10.1 s is square if and only if there exists an integer n such that $s = n^2$.

Exercise 10.2 Use this definition to show that 4 is a square number and so is $x^2 + 4x + 4$, given x is an integer.

The absolute value of a real number is defined as its distance from zero. The formal definition that we will use for absolute value of a real number x is

Definition 10.2 $|x| = \begin{cases} x & \text{if } x \geq 0 \\ -x & \text{if } x < 0 \end{cases}$.

Exercise 10.3 Re-write each expression so that it no longer has an absolute value in it. Assume that x is a real number.

$|-.5|$

$|x^2 + 4x + 4|$

$|x^2 + 6x + 10|$

$|-x^2 + 2x - 1|$

$|-x|$ given $x < 0$

10.2 Inequality properties

At this point it will be useful to review some properties of inequalities and absolute value. If you have two inequalities $A < B$ and $C < D$ then it is true that $A + C < B + D$. For example, we know $3 < 4$ and $4 < 5$. Adding these inequalities together yields $7 < 9$, which is true.

Exercise 10.4 Prove that the same property does not hold for subtracting inequalities.

Property 10.1 Inequalities facing the same direction can be added together but they cannot be subtracted.

Property 10.2 For all real numbers x, $-|x| \leq x \leq |x|$.

Property 10.3 For all real numbers x and y, $|xy| = |x| \cdot |y|$.

Exercise 10.5 Explain why Property 2 is true. In your explanation include examples of both negative and positive values. We will prove this in Chapter 18.

Exercise 10.6 We will prove Property 3 when we get to Chapter 12 because it uses a proof technique that we have not learned yet. For now, try out some examples and explain why you think it is true.

10.3 Proofs

Exercise 10.7 What do you notice and wonder about the following image?

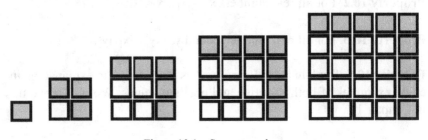

Figure 10.1. Square numbers

_____

_____

_____

_____

Exercise 10.8 Make a conjecture of what would result if you subtract two consecutive square numbers. Give some examples to back up your conjecture and then prove it.

_____

_____

Example 10.1 Prove or disprove: $\forall \, a, b \in \mathbb{Z}$, if $a | b$ then $|a| \leq |b|$.

We know that $a|b$, which is equivalent to saying $b = ak$, where k is an integer. Let's try an example to see if we can prove the result. Let $a = 2$ and $b = 4$. It is true that $2|4$. We can see that $|2| \leq |4|$, so the result will hold. It seems like this will always hold when a and b are positive values because in that case if we know $a|b$, b would have to be larger than a. Let's look at negative numbers. Let $a = -2$ and $b = 4$. It is true that $-2|4$ because $4 = -2(-2)$. $|-2| \leq |4|$ also holds true. It also appears that negative numbers will make this statement true. What about 0? If $a = 0$ then $b = 0k = 0$ and it is true that $0 \leq 0$. If $b = 0$ then $0 = ak$. The only way this can happen is if $a = 0$ or $k = 0$. If $a = 0$ we are back to the situation where $0 \leq 0$, so that is not a problem. If $k = 0$ then a could be any integer. For example, say $a = 5$. Then 5 is not less than or equal to 0. That is a counterexample that disproves this statement.

Disproof: Let $a = 5$ and $b = 0$. It is true that $5|0$ because $0 = 5(0)$. The conclusion does not hold since 5 is not less than or equal to 0. $\square$

In Chapter 7 we went over the idea of a lemma. Sometimes when you are trying to prove a theorem you realize that it would be helpful to have a result that you haven't yet proven. Rather than writing a proof within a proof, it makes more sense to go to the side and write out this smaller proof. We call this a lemma. Let's look at an example to see how this idea can be used.

Example 10.2 Prove for all integers n, if n is even then n^2 is even.

Let us start by putting the given into a more useable form; $n = 2k$, where k is an integer. We want to prove that $n^2 = 2(\text{integer})$. While we can prove this directly as usual, it will be quicker and more efficient to use a lemma. We can think of n^2 as $n \cdot n$. Back in Homework 8.1 we proved that an even number times an even number is even. Since we proved this already you do not have to prove it again; we can simply use the result. Thus, we are using it as a lemma. We are given n is even, which means that $n \cdot n = \text{even} \cdot \text{even} = \text{even}$ by Homework 8.1. This proves what we want, that n^2 is even.

Proof.

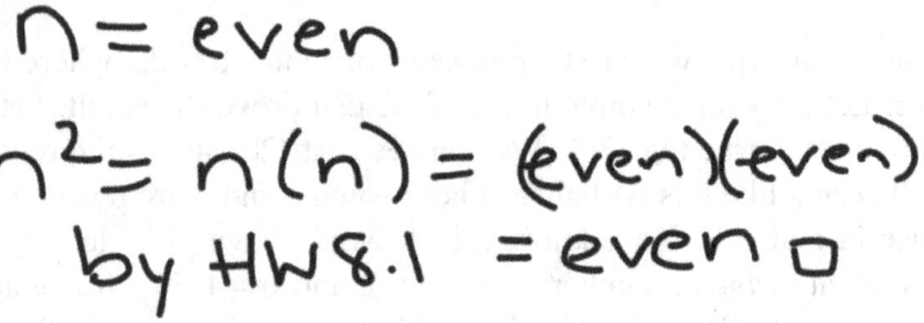

$$n = even$$
$$n^2 = n(n) = (even)(even)$$
$$by\ HW\ 8.1 = even \quad \square$$

Figure 10.2.   Proof for square of even number is even

In Appendix E you can find a list of selected proofs from all chapters that may be useful as lemmas. We also went over a corollary in Chapter 7. We said a corollary is when you can get an additional result without doing too much extra work. Let's look at an example to see how this idea can be used.

Example 10.3 First we will prove for all integers n and $k > 1$, if $2 | n$ then $2 | n^k$.

We are given $2 | n$, which translates to $n = 2j$, where j is an integer. We want to prove that $n^k = 2(\text{integer})$. If we start with the given, we should be able to raise both sides to the kth power and get what we want to prove. Let's try it.

Proof. $n^k = (2j)^k = 2^k j^k = 2(2^{k-1} j^k) = 2(\text{integer})$. $\square$

Note: Since we are given $k > 1$ we know that $k - 1 > 0$, making 2^{k-1} a positive power of 2.

Suppose we are also asked to prove for all integers n, if $2 | n$ then $2 | n^2$. We could go through the same proof with a 2 where the k was before, but another option is to not repeat all the work that we just did. We could simply say as a corollary to Example 10.3 with $k = 2$ this will be true. We proved it is true for all integers k greater than 1, so it will be true when $k = 2$. Any time you have already proved a more general case already you can use a corollary to prove the more specific case.

Exercise 10.9 Complete the following. Use a copy of the proof framework or list your steps following the same format.

(1) Prove or disprove: One more than the product of consecutive odd numbers is a perfect square.

(2) Prove or disprove: $\forall n \in \mathbb{Z}$, $4(n^2 + n + 1) - 3n^2$ is a square number.

(3) Prove or disprove: $\forall a, b \in \mathbb{Z}$, if $a \mid b$ then $|a|$ divides $|b|$.

(4) Prove or disprove: $\forall x, y \in \mathbb{R}$, $|x - y| \le |x| - |y|$.

Exercise 10.10 Summarize what we learned in the direct proofs chapters. What questions or concerns do you have about the material you have learned so far?

Chapter 10 Homework

Fill in your homework log as you work on the following problems.

(1) Prove or disprove: $\forall x \in \mathbb{Z}$, $4x^2 + 4x + 1$ is a square number.
(2) Prove or disprove: The product of any rational number and any integer is rational.
(3) Prove or disprove: Every integer divides 0.
(4) Find all values of n where $2n^2 + 7n + 3$ is prime.
(5) Prove or disprove: $\forall x \in \mathbb{R}$, $|x| = 0$ if and only if $x = 0$.
(6) Use the result of Homework 8.3 [For all integers a,b, x, and y, if $a|b$ then $a|(ax + by)$] to prove the corollary: For all integers a and b, if $a|b$ then $a|(2a + 3b)$.

Chapter 11

Direct Proofs (GCD & Relatively Prime)

11.1 Definitions

Definition 11.1 If a and b are nonzero integers then the greatest common divisor of a and b, $\gcd(a,b)$, is the largest integer d such that $d|a$ and $d|b$.

Exercise 11.1 What is $\gcd(88,448)$? What about $\gcd(12x,x)$, given x is an integer?

Exercise 11.2 Is there always a greatest common divisor for two integers? What is the smallest possible gcd?

_____

_____

_____

_____

Definition 11.2 Integers a and b are relatively prime if $\gcd(a,b) = 1$.

Be careful not to confuse the term relatively prime with the term prime.

Exercise 11.3 Show that 15 and 28 are relatively prime.

11.2 Proofs

Example 11.1 Find all possible values for $\gcd(a+2,2)$, where a is an integer.

With gcd proofs it is generally helpful to give your gcd a name so that it is easy to refer to. Let $\gcd(a+2,2)=d$, where d is an integer ≥ 1. Using the definition, we know that d is the largest integer such that $d|(a+2)$ and $d|2$. Since $d|2$ we know that $d=-2,-1,1,$ or 2. But we also know that $d\geq 1$. That means the only possibilities left for d are 1 or 2. The answer will depend on whether a is even or odd. If a is even then $a+2$ will be even, which means that $2|(a+2)$ and $\gcd(a+2,2)=2$. When a is odd then 2 does not divide $a+2$ so $\gcd(a+2,2)=1$.

Proof.

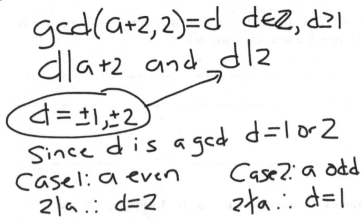

Figure 11.1. Proof for $\gcd(a+2,2)$

Example 11.2 Prove for any prime p and integers a and b, if $\gcd(a, p^3) = p^2$ and $\gcd(b, p^3) = p$ then $\gcd(a + b, p^3) = p$.

Proof. We know that p is prime so its only factors are 1 and itself. We also know that p^2 is the largest value that divides both a and p^3. This means that a has 2 copies of p as factors. We also know that p is the largest value that divides both b and p^3. This means that b has only 1 copy of p as a factor. Since b has only 1 factor of p while a has 2 factors of p, a + b will only have 1 factor of p. Thus, $\gcd(a + b, p^3) = p$. $\square$

We have now defined the following terms: even number, odd number, one integer dividing another integer, rational number, prime number, composite number, square number, absolute value, greatest common divisor, and relatively prime. These are terms that you need to know in order to be successful for the remainder of this class. As we go through examples you will naturally start to remember the definitions, but as we saw in Chapter 6, you should actively study them. One way to accomplish this is to create a set of flash cards for each of these terms. An online set of flash cards has been created on Quizlet. You can modify these to make them more personal. On Quizlet you can play different games with these cards. View the cards here: https://quizlet.com/374230056/proofs-flash-cards/.

Exercise 11.4 Complete the following. Use a copy of the proof framework or list your steps following the same format.

(1) Prove or disprove: For all prime numbers p and integers a, if p|a then $\gcd(p, a) = p$.
(2) Prove or disprove: For all prime numbers p and integers a, if p does not divide a then a and p are relatively prime.
(3) Prove or disprove: $\forall m \in \mathbb{Z}$, $\gcd(0, m) = m$.

Exercise 11.5 Write a reflection on what you learned from any of the mistakes you or a classmate made on any of the proofs you did today.

_____

_____

_____

Chapter 11 Homework

Fill in your homework log as you work on the following problems.

(1) Prove or disprove: The square of any rational number is rational.
(2) Prove or disprove: $\forall a, b \in \mathbb{Z}$, if $a \mid 7b$ then $a \mid b$.
(3) Prove or disprove: $\forall r, s \in \mathbb{Z}$, if r and s are relatively prime then $r + s$ and s are relatively prime.
(4) Prove or disprove: For all non-zero integers a and b, if x and y are any integers then $\gcd(a, b)$ divides $(ax + by)$.
(5) Pick some of the terms below and make a visual representation of them as a way of helping you remember what they mean.
    (a) Negation of for all/there exist statements
    (b) Negations of if-then statements
    (c) Even/Odd
    (d) Rational
    (e) Divides
    (f) Prime/Composite
    (g) Absolute Value

Example: A poster similar to this was hanging up in my 8th grade math classroom and I still visualize it to remember that a vertical line has an undefined slope.

Figure 11.2. No slope

Chapter 12

Proof by Division into Cases

12.1 Proof mistakes

Before moving on to the new topic we will review some common mistakes made on proofs.

Mistake 1: Giving an example as a proof.

Exercise 12.1 What is wrong about the following "proof"?

Prove any even number plus any even number is even.

Proof. 2 is even and 4 is even. $2 + 4 = 6$ is also even. $\square$

_____

_____

_____

_____

Mistake 2: Using what you want to prove.

Exercise 12.2 What is wrong about the following "proof"?

Prove: $\forall n \in \mathbb{Z}, 4(n^2 + n + 1) - 3n^2$ is a square number.

Proof.

$$4(n^2+n+1)-3n^2=k^2 \quad k \in \mathbb{Z}$$
$$4n^2+4n+4-3n^2=k^2$$
$$n^2+4n+4=k^2$$
$$(n+2)^2=k^2$$
$$n+2=k \quad \square$$

Figure 12.1. Exercise 12.2 proof

_____

_____

_____

_____

Mistake 3: Writing what you want to prove.

Exercise 12.3 What is wrong about the following "proof"?

Prove for all integers a and b, if a|b then a|6b.

Proof.

$$b=ak$$
$$6b=aj \quad k,j \in \mathbb{Z}$$
$$6b=6(ak)=a(6k)$$
$$=a(int) \quad \square$$

Figure 12.2. Exercise 12.3 proof

Mistake 4: Introducing a letter but not saying what it represents.

Exercise 12.4 What is wrong about the following "proof"?

Prove if n is a positive power of 2 then $n + 8$ is composite.

Proof.

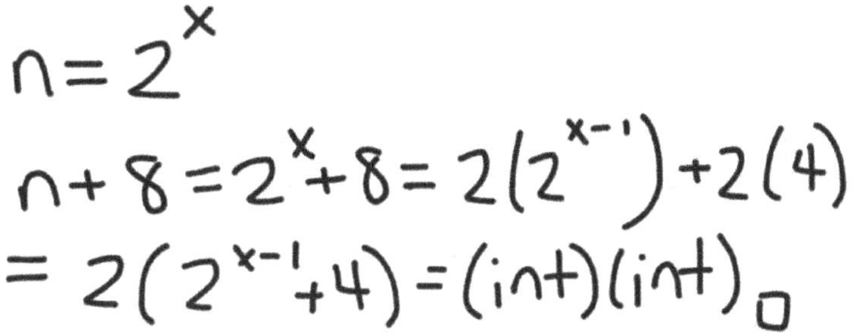

$$n = 2^x$$
$$n + 8 = 2^x + 8 = 2(2^{x-1}) + 2(4)$$
$$= 2(2^{x-1} + 4) = (int)(int) \quad \square$$

Figure 12.3. Exercise 12.4 proof

Mistake 5: Saying something is an integer just because you want it to be.

Exercise 12.5 What is wrong about the following "proof"?

Prove for all integers a and b, if a|6b then a|b.

Proof.

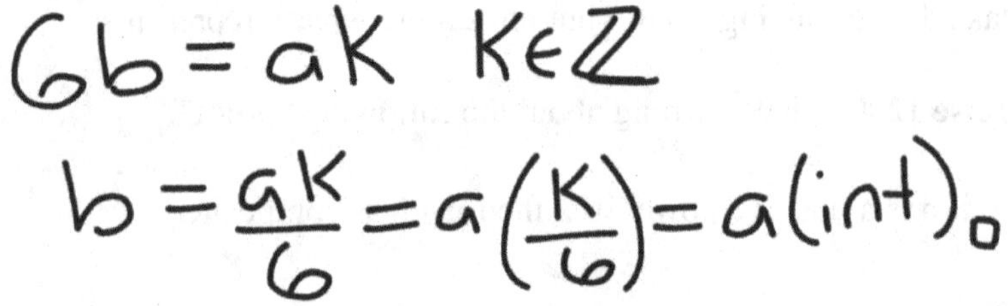

$$6b = aK \quad K \in \mathbb{Z}$$

$$b = \frac{aK}{6} = a\left(\frac{K}{6}\right) = a(int).\quad _\square$$

Figure 12.4. Exercise 12.5 proof

_____

_____

_____

_____

Mistake 6: Using the same letter to stand for two different things.

Exercise 12.6 What is wrong about the following "proof"?

Prove the difference of any two rational numbers is rational.
Prove: $\forall x, y \in \mathbb{R}$, if x and y are rational then $x - y$ is rational.

Proof.

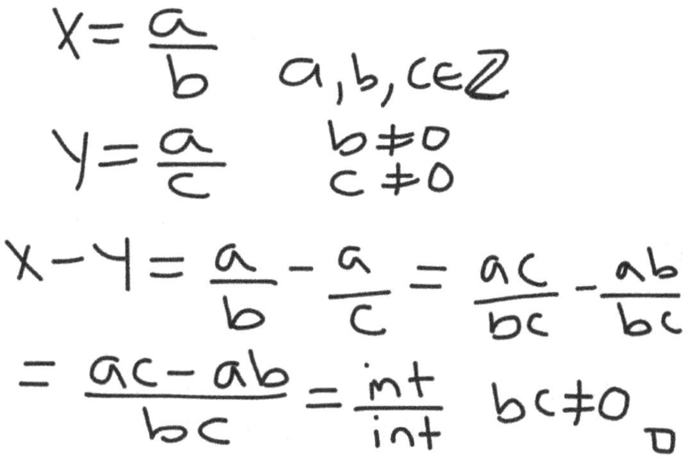

$$X = \frac{a}{b} \qquad a, b, c \in \mathbb{Z}$$

$$Y = \frac{a}{c} \qquad b \neq 0$$
$$\qquad\qquad c \neq 0$$

$$X - Y = \frac{a}{b} - \frac{a}{c} = \frac{ac}{bc} - \frac{ab}{bc}$$

$$= \frac{ac - ab}{bc} = \frac{int}{int} \quad bc \neq 0 \qquad \square$$

Figure 12.5. Exercise 12.6 proof

12.2 Proof by division into cases

So far we have done all proofs directly. Starting with this chapter, we will begin to learn new proof techniques. In the last four chapters we learned new vocabulary terms. From here on instead of learning new terms, we will apply those same definitions to new proof techniques. The first new proof technique is called division into cases. In this chapter we will look at two definitions that naturally fit into this category: absolute value and even/odd numbers. The definition of absolute value is given as a piecewise function with two cases, so it should be no surprise that many proofs that have an absolute value can be done by splitting into cases. We will look at two such examples.

Example 12.1 Prove: $\forall\, x, a \in \mathbb{R}$, $|x| < a$ *if and only if* $a < x < a$.

First let's re-write this as two conditional statements and prove each one separately.

Prove: $\forall\, x, a \in \mathbb{R}$, *if* $|x| < a$ *then* $-a < x < a$.

Proof. The given is that $|x| < a$. I will apply the definition of absolute value and split this into two cases.

Case 1: If $x \geq 0$ then we know $|x| = x$, which turns the given into $x < a$. Since x is positive and less than a, that means a is also positive, making $-a$ negative. A negative number is always less than a positive number so we also know that $-a < x$. Putting these together yields $-a < x < a$.

Case 2: If $x < 0$ then we know $|x| = -x$, which means that $-x < a$. Multiplying both sides by -1 yields $x < -a$. Since x is negative and greater than $-a$, that means $-a$ is also negative, making a positive. A negative number is always less than a positive number so we also know that $x < a$. Putting these together yields $-a < x < a$. $\square$

Prove: $\forall\, x, a \in \mathbb{R}$, *if* $-a < x < a$ then $|x| < a$.

Proof. In this direction we know that $-a < x < a$. Again, let's use two cases.

Case 1: If $x \geq 0$ then we know $|x| = x$. Since $x < a$ is given we can substitute to get $|x| < a$.

Case 2: If $x < 0$ then we know $|x| = -x$. Since $-a < x$ is given we can substitute to get $-a < -|x|$. Multiplying both sides by -1 yields $|x| < a$. $\square$

In general, when you see an absolute value in a proof you will want to get rid of the absolute value. The reason for this is that absolute values do not behave nicely. Try this next exercise to see why.

Exercise 12.7 For any real numbers x and y, does $|x+y| = |x| + |y|$? Try some examples out.

This would hold true if you chose x and y to both be positive or both negative. But if you choose one to be positive and the other to be negative then you have a problem. So sometimes the equality holds and sometimes it doesn't. Let's try to relax the condition to get a statement that is always true.

Exercise 12.8 Use your examples from Exercise 12.7 or add more examples to figure out how to modify the statement $|x+y| = |x| + |y|$ to make it always true.

Hopefully you were able to get the following statement. It is so often used that it gets its own name: The Triangle Inequality.

Theorem 12.1 (Triangle Inequality) $\forall x, y \in \mathbb{R}, \ |x+y| \leq |x| + |y|$.

Example 12.2 Prove the Triangle Inequality.

Proof. The proof that we will do together actually doesn't use cases. We will use some of the absolute value properties discussed in Chapter 10. First, we will put what we want to prove into a more useable form. In Example 12.1 we proved that if $|x| < a$ then $-a < x < a$. We can re-write this letting $x = x + y$ and $a = |x| + |y|$.

$$-(|x| + |y|) < x + y < |x| + |y|. \tag{12.1}$$

In Chapter 10 we saw Property 10.2: For all real numbers x, $-|x| \le x \le |x|$. Similarly, we can say that $-|y| \le y \le |y|$. In Property 10.1 we saw that adding two inequalities that go in the same direction is allowed. Adding these together yields

$$-|x| + -|y| \le x + y \le |x| + |y|. \qquad (12.2)$$

Factoring out a negative on the left side gives us what we wanted to prove. □

Exercise 12.9 Now prove The Triangle Inequality using cases.

As we said earlier, it is generally a good idea to get rid of the absolute value to make the proof easier to deal with. Since the definition of absolute value is a piece-wise function, that will help us decide on what cases to use. Whatever is located within the absolute value is what you apply the definition to. For example, if the proof involves a term of the form $|x + 4|$ then your cases would be $x + 4 \geq 0$ and $x + 4 < 0$.

Exercise 12.10 Complete the following. Use a copy of the proof framework or list your steps following the same format.

(1) Prove or disprove: $\forall a, b \in \mathbb{Z}$, if ab is even then a or b is even.
(2) Prove or disprove: For all real numbers x and y, $|xy| = |x| \cdot |y|$.
(3) Prove or disprove: The product of any two consecutive integers is even.
(4) Prove or disprove: If n is an integer then $3n^2 + n + 14$ is even.

Exercise 12.11 Write a reflection on what you learned from any of the mistakes you or a classmate made on any of the proofs you did today.

_____

_____

_____

_____

Chapter 12 Homework

Fill in your homework log as you work on the following problems.

(1) Prove for all integers n, $n^3 + 3n$ is even.

(2) Prove for all real numbers x, $x + |x - 7| \geq 7$.

(3) Prove or disprove: $\forall x, y \in \mathbb{R}$, $|x - y| = |y - x|$.

(4) Prove or disprove: $\forall x, y \in \mathbb{R}$, $x = y$ if and only if $xy = \frac{(x+y)^2}{4}$.

(5) Prove or disprove: For all integers $m > 3$, $m^2 - 4$ is composite.

(6) Prove or disprove: $\forall x, y \in \mathbb{Z}$, if $x^2 + y^2$ is even then $x + y$ is even.

Chapter 13

Proof by Division into Cases (Quotient Remainder Theorem)

13.1 Quotient remainder theorem

Exercise 13.1 Your professor is going to call out a number and you have 10 seconds to form groups of that size. After each round think about if there were any people left over who were not able to form a group and why. Then fill out the table.

Table 13.1. Quotients and remainders

Group size (d)	# of groups	Remainder (r)
2		
3		
4		
5		
6		

Exercise 13.2 Think of the class size as a number n, the group size as a divisor d, and how many people left over as the remainder r. Write an equation that relates all three of these values.

Exercise 13.3 What are the possible remainders in a class size of 2 people? What about a class size of 3 people? 4 people? n people?

Table 13.2. Remainders

Class size	Possible remainders
2	
3	
4	
n	

This example of forming groups illustrates what is known as the quotient remainder theorem.

Theorem 13.1 (Quotient Remainder Theorem) $\forall n \in \mathbb{Z}, \exists d, k, r \in \mathbb{Z}$ such that $n = dk + r$, where $0 \leq r \leq n - 1$.

We saw that the possible remainders of a number n go from 0 to one less than the number. You may have noticed that the remainders then start to repeat. For example,

$$24 = 3(8) + 0$$
$$25 = 3(8) + 1$$
$$26 = 3(8) + 2$$
$$27 = 3(8) + 3 = 3(9) + 0$$

Rather than saying 27 has a remainder of 3 we bring that 3 over to the divisor. You also probably wouldn't say that something has a negative remainder. But we can see what that would look like and do the same kind of manipulation to turn the remainder positive. For example, $25 = 3(9) - 2 = 3(8) + 3 - 2 = 3(8) + 1$. We took one copy of 3 out of the 9 copies and then combined it with the -2 to show that this is equivalent to a remainder of positive 1. We won't go into too much depth on this topic but just to give you a quick idea of how it works in general think of a clock; in fact this idea of manipulating remainders is sometimes called clock arithmetic. This clock has the remainders when dividing by 3. When dividing a number by 3, as the number increases by 1, the remainder will move one position to the right. We can also use this clock idea for negative remainders. The only difference is you would go left rather than right. In this way you can see that a remainder of -1 is equivalent to a remainder of 2.

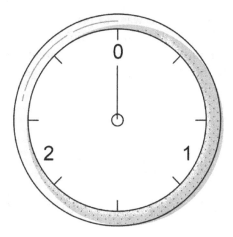

Figure 13.1. Clock arithmetic

Exercise 13.4 When dividing by 9, what is a remainder of -1 equivalent to?

13.2 Proofs

The Quotient Remainder Theorem lends itself to proofs that use division into cases.

Example 13.1 Prove: $\forall n \in \mathbb{Z}, 2|n(n+1)$.

First, let's notice that this proof would not work directly. We are not given any information beyond n is an integer, so there is no way to know if 2 divides $n(n+1)$. We are asked to prove something about divisibility by 2 so that will guide how many cases we choose to begin with. Our number is n. We will use the quotient remainder theorem to get two cases. Case 1: n is 2(integer) and Case 2: n is 2(integer)+1. In Case 1 let the integer be called k. We want to prove that $n(n+1) = 2$(integer). Then $n(n+1) = 2k(2k+1)$. We could multiply this out but to save time and effort you may notice that putting proper parentheses will get this in the form we want, $n(n+1) = 2[k(2k+1)] = 2$(integer). Here is Case 1 written out.

Proof.

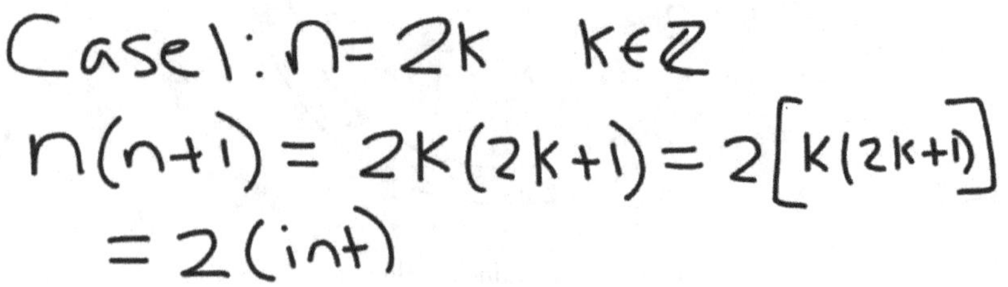

$$\text{Case 1: } n = 2k \quad k \in \mathbb{Z}$$
$$n(n+1) = 2k(2k+1) = 2\left[k(2k+1)\right]$$
$$= 2(\text{int})$$

Figure 13.2.   Example 13.1 Case 1 proof

Exercise 13.5 For Case 2 we have $n = 2$(integer)+1. Why is it okay to call this integer k like we did in Case 1?

_____

_____

_____

Let's complete Case 2. We have $n(n+1) = (2k+1)(2k+2)$. Again, we could multiply these out, but it will be more efficient to factor a 2 out of the second part to get $2[(2k+1)(k+1)] = 2(\text{integer})$. The quotient remainder theorem tells us that every integer can be expressed in one of the two cases (i.e. every integer is either even or odd) and we have proved that regardless of which case the number satisfies, 2 will divide $n(n+1)$. Here is Case 2 written out.

$$\text{Case } 2: n = 2k+1 \quad k \in \mathbb{Z}$$
$$n(n+1) = (2k+1)(2k+2)$$
$$= 2[(2k+1)(k+1)] = 2(\text{int}) \qquad \square$$

Figure 13.3. Example 13.1 Case 2 proof

Example 13.2 Prove: The square of any odd integer has a remainder of 1 when divided by 8.

Let's begin by putting this into formal notation.
$\forall n \in \mathbb{Z}$, if n is odd then $n^2 = 8(\text{integer}) + 1$.

Exercise 13.6 Show that this cannot be proven directly.

Exercise 13.7 How many cases should we use to prove this statement?

In Exercise 13.7 you may have answered 8 cases because the proof has to do with divisibility by 8. That is a correct answer, but the good news is that sometimes we can also do it in fewer. This is problem-specific. In this one we need to be able to factor an 8 out of the square of the number we start with. It makes sense that 2 cases won't work because $2^2 = 4$ and that is too small to be able to factor out an 8. And 3 also won't work because 8 does not have a factor of 3 in it. However, 4 will work because $4^2 = 16$ and it is possible to factor an 8 out of 16. For that reason, we will start with 4 cases: $n = 4(\text{int})$, $4(\text{int})+1$, $4(\text{int})+2$, or $4(\text{int})+3$. The extra good news for this problem is that we also know that n is odd so we can eliminate the cases where n is even; namely, $4(\text{int})$ and $4(\text{int})+2$.

Proof.

$$\text{Case 1: } n = 4k+1$$
$$n^2 = (4k+1)^2 = 16k^2 + 8k + 1 = 8(8k^2+k)+1$$
$$= 8(\text{int})+1$$
$$\text{Case 2: } n = 4k+3$$
$$n^2 = (4k+3)^2 = 16k^2 + 24k + 9$$
$$= 16k^2 + 24k + 8 + 1 = 8(8k^2+3k+1)+1$$
$$= 8(\text{int})+1$$

Figure 13.4. Example 13.2 proof

Every odd integer falls into one of these two cases and we have proved $n^2 = 8(\text{integer})+1$ for both of them. $\square$

Exercise 13.8 Complete the following. Use a copy of the proof framework or list your steps following the same format.

(1) Prove or disprove: $\forall n \in \mathbb{Z}$, if n is not divisible by 5 then n^2 has a remainder of 1 or 4 when divided by 5.

(2) Prove or disprove: *If* $p > 3$ with p and $p + 2$ prime then $6|(p + 1)$.

(3) $\forall n \in \mathbb{Z}$, if $3|(n^2 + 8n)$ what are the possible remainders when n is divided by 3?

(4) Prove or disprove: For all even integers m, 4 does not divide $2m - 2$.

Exercise 13.9 Write a reflection on what you learned from any of the mistakes you or a classmate made on any of the proofs you did today.

Chapter 13 Homework

Fill in your homework log as you work on the following problems.

(1) Prove that the product of any two consecutive integers has the form $3k$ or $3k + 2$ for some integer k.

(2) Given p is prime, show $p + 2$ and $p + 4$ aren't also prime unless $p = 3$.

(3) Prove for all integers m, if m is not divisible by 3 then $m^2 + 2$ is divisible by 3.

(4) Prove: Every integer that is a perfect cube is either a multiple of 9, or 1 more, or 1 less than a multiple of 9.

(5) Prove or disprove: $\forall\, r, s \in \mathbb{Z}$, if $r|s$ and $3|r$ then $3|s$.

Chapter 14

Forward-Backward Proofs

14.1 Proofs

In Forward-Backward proofs you start looking at what you're trying to prove and see how to get to it from an obviously true statement or a point where it is obvious how to prove it. The proof is your steps reversed; start with the true statement and get to what you're trying to prove. You end the scratch work when you come to an obviously true statement, either a given or something you know is always true, like (number)$^2 \geq 0$. This technique only works if each step you do is reversible.

Example 14.1 Prove: $\forall x, y \in \mathbb{R}, \frac{(x+y)^2}{4} \geq xy$.

Scratch work.

$$\frac{(x+y)^2}{4} \geq xy$$

$$(x+y)^2 \geq 4xy$$

$$x^2 + 2xy + y^2 \geq 4xy$$

$$x^2 - 2xy + y^2 \geq 0$$

$$(x-y)^2 \geq 0$$

Figure 14.1. Example 14.1 scratch work

This is true for any x and y so we can stop there and begin the proof by following the steps in reverse.

Proof.

$$(x-y)^2 \geq 0$$

$$x^2 - 2xy + y^2 \geq 0$$

$$x^2 + 2xy + y^2 \geq 4xy$$

$$(x+y)^2 \geq 4xy$$

$$\frac{(x+y)^2}{4} \geq xy \quad \square$$

Figure 14.2. Example 14.1 proof

Exercise 14.1 Why do you start with what you want to prove in a Forward-Backward proof but this is not allowed in a direct proof?

_____

_____

_____

_____

14.2 Concise proofs

Now that we have completed many proofs, let's discuss what makes a good proof. To start, a proof has to be valid. There is never one exact right way to complete a proof. Rather, there are often many different ways that are correct and valid. But you also may have noticed that even though a proof is valid sometimes it isn't concise. A valid, concise proof is the ultimate goal. Why prove something in 10 lines when you can prove it in 5?

Exercise 14.2 Take a look at this correct proof. Make sure you understand it and then try to create a different proof that is more concise.

Prove: For all positive integers n, $n^2 + 3n + 2$ is not prime.

Proof.

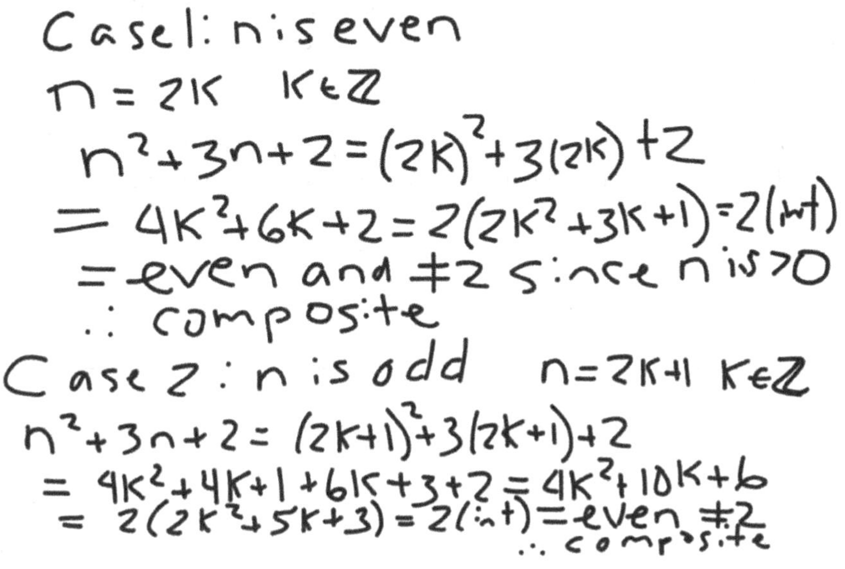

$$\text{Case I: } n \text{ is even}$$
$$n = 2K \quad K \in \mathbb{Z}$$
$$n^2 + 3n + 2 = (2K)^2 + 3(2K) + 2$$
$$= 4K^2 + 6K + 2 = 2(2K^2 + 3K + 1) = 2(M)$$
$$= \text{even and } \neq 2 \text{ since } n \text{ is} > 0$$
$$\therefore \text{ composite}$$
$$\text{Case 2: } n \text{ is odd} \quad n = 2K+1 \quad K \in \mathbb{Z}$$
$$n^2 + 3n + 2 = (2K+1)^2 + 3(2K+1) + 2$$
$$= 4K^2 + 4K + 1 + 6K + 3 + 2 = 4K^2 + 10K + 6$$
$$= 2(2K^2 + 5K + 3) = 2(\ldots) = \text{even} \neq 2$$
$$\therefore \text{ composite}$$

Figure 14.3. Exercise 14.2 proof

Exercise 14.3 So far we have learned direct proofs, proof by division into cases, and Forward-Backward proofs. For each question below first decide what method you would likely use and then prove each statement.

(1) $\forall x \in \mathbb{R}$, If $x > 10$ then $x^2 + 40 > 14x$.
(2) $\forall n \in \mathbb{Z}$, $n^2 + n + 1$ is odd.
(3) If x and y are two different positive real numbers then $\frac{x+y}{2} > \sqrt{xy}$.
(4) For all integers $n > 1$, 2^n is composite.

Exercise 14.4 At the end of the book in Appendix F you will find a table to fill in explaining each proof method that we will cover. Fill in the first three that we have learned so far with an explanation as well as an example.

Exercise 14.5 Write a reflection on what you learned from any of the mistakes you or a classmate made on any of the proofs you did today.

_____

_____

_____

_____

Chapter 14 Homework

Fill in your homework log as you work on the following problems.

(1) Prove or disprove: $\forall x \in \mathbb{R}$, if $x > 11$ then $x^2 > 9x + 22$.

(2) Prove or disprove: $\forall x \in \mathbb{R}$ with $x \neq -1$ and $x \neq -2$, $\dfrac{x}{x+1} < \dfrac{x+1}{x+2}$.

(3) Prove or disprove: $\forall n \in \mathbb{R}$, if $3|n$ then $2n^2 - 3n$ is divisible by 9.

(4) Follow the steps for a Forward-Backward proof on the problem below and explain why this method doesn't work. Then disprove the statement $\forall x \in \mathbb{R}$, $x^2 + 6 > 5x$.

(5) Prove or disprove: Any rational number divided by 5 is still rational.

(6) Explain what is good about the following proof, what is incorrect, and how it can be fixed to be correct. Include a correct proof in your explanation. Prove for all integers m, if $m^2 + 2$ is even then m is even.

$$m = 2k \quad k \in \mathbb{Z}$$
$$m^2 + 2 = (2k)^2 + 2 = 4k^2 + 2$$
$$= 2(2k^2 + 1) = 2(\cdot n +) = even \checkmark$$

Figure 14.4. Homework 14.6 proof

Chapter 15

Proof by Contraposition

15.1 Contrapositive review

Exercise 15.1 Explain why the following statement can't be proven directly or by cases: $\forall a, b \in \mathbb{R}$, if ab is irrational then a is irrational or b is irrational.

_____

_____

_____

_____

_____

_____

In our logic chapter we saw that a conditional statement and its contrapositive have the same truth value. This is useful to know because sometimes a statement is easier to prove in its contrapositive form. We will call these types of proofs "proof by contraposition." These types of proofs are called indirect because we need to manipulate the original statement before completing the proof. In the next chapter we will learn another type of indirect proof.

Exercise 15.2 Write the contrapositive of the following statements.

(1) $\forall a, b \in \mathbb{R}$, if ab is irrational then a is irrational or b is irrational.

(2) $\forall n \in \mathbb{Z}$, if n is even then n^2 is even.

15.2   Proof by contraposition

Example 15.1 Prove: $\forall a, b \in \mathbb{R}$, if ab is irrational then a is irrational or b is irrational.

We saw that this can't be proved directly so the first step is to write the contrapositive.

Contrapositive: $\forall a, b \in \mathbb{R}$, if a is rational and b is rational then ab is rational.

Putting the givens into useable form gives $a = \frac{m}{n}, b = \frac{j}{k}$ where m, n, j, k are integers and $n \neq 0$ and $k \neq 0$. Then $ab = \frac{mj}{nk}$. We know mj and nk are integers and $nk \neq 0$ since n and k separately are non-zero. This fits the definition for ab being rational. This means that the original statement is also true since the contrapositive and the original statement have the same truth value. Here is the proof written out.

Proof.

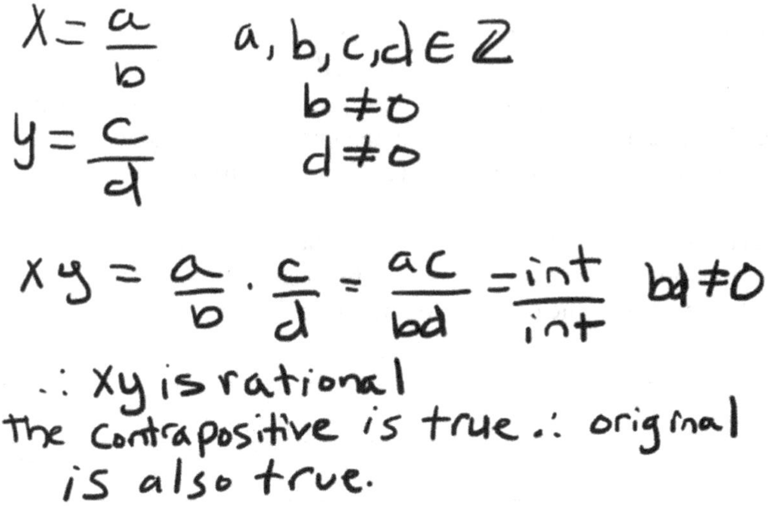

$$X = \frac{a}{b} \qquad a, b, c, d \in \mathbb{Z}$$
$$\qquad\qquad b \neq 0$$
$$y = \frac{c}{d} \qquad\qquad d \neq 0$$

$$Xy = \frac{a}{b} \cdot \frac{c}{d} = \frac{ac}{bd} = \frac{int}{int} \qquad bd \neq 0$$

∴ Xy is rational
The contrapositive is true ∴ original
is also true.

Figure 15.1. Example 15.1 proof

Example 15.2 Prove: $\forall n \in \mathbb{Z}$, n is even if and only if n^2 is even.

First we will split this up into two conditional statements and prove each separately.

Prove: $\forall n \in \mathbb{Z}$, if n is even then n^2 is even.

Proof. This proof can be done directly. Since n is even it can be expressed as $n = 2k$, where k is an integer. Then $n^2 = (2k)^2 = 4k^2 = 2(2k^2) = 2(\text{integer}) = \text{even}$. $\square$

Prove: $\forall n \in \mathbb{Z}$, if n^2 is even then n is even.

Proof. This direction can't be proven directly because you would have to start with n^2 being even. If you try to solve for n it would involve a square root and at that point you are getting away from the set of integers. The first step is to write the contrapositive.

Contrapositive: $\forall\, n \in \mathbb{Z}$, if n is odd then n^2 is even.

Now the proof works similarly to the first direction. Let $n = 2k + 1$, where k is an integer. Then $n^2 = (2k+1)^2 = 4k^2 + 4k + 1 = 2(2k^2 + 2k) + 1 = 2(\text{integer}) + 1 = \text{odd}$. We have now proved that the contrapositive is true. This means that the original statement is also true since the contrapositive and the original statement have the same truth value. $\square$

Exercise 15.3 Look at the examples we have done so far for proof by contraposition. Describe when it will be easier to prove the contrapositive statement over the original conditional statement.

_____

_____

_____

_____

_____

_____

Exercise 15.4 Complete the following. Use a copy of the proof framework or list your steps following the same format.

(1) Prove: $\forall\, a, b, c, d \in \mathbb{Z}$, if $(ax+b)(cx+d) = 1$ then $a \neq 0$ or $a \neq 0$.
(2) Prove: $\forall\, a \in \mathbb{Z}$, 5 does not divide a^4 *iff* 5 does not divide a.
(3) Prove: $\forall\, k \in \mathbb{Z}$, if $3k + 1$ is even then k is odd.
(4) Prove: $\forall\, a, b \in \mathbb{Z}$, if $a + b \geq 16$ then $a \geq 8$ or $b \geq 8$.

Exercise 15.5 Write a reflection on what you learned from any of the mistakes you or a classmate made on any of the proofs you did today.

_____

_____

_____

_____

_____

_____

Exercise 15.6 Fill in the "Proof Methods" table explaining how proof by contraposition works as well as providing an example.

Chapter 15 Homework

Fill in your homework log as you work on the following problems.

(1) Prove: $\forall x \in \mathbb{Z}$, if x^3 is even then x is even.

(2) Prove: $\forall n \in \mathbb{Z}$, if $1 + m^7$ is even then m is odd.

(3) Prove for all integers x, $11x - 7$ is even if and only if x is odd.

(4) Answer each part.

    (a) Come up with a counterexample to disprove: For all integers a and b, if $a|9b$ then $a|9$ or $a|b$.

    (b) Prove: For all integers a and b, if $a|5b$ then $a|5$ or $a|b$.

    (c) Explain the difference between problems a and b (why a is false and b is true?).

(5) Prove or disprove: $\forall x, y \in \mathbb{R}, \; |x - y| \leq |x| + |y|$.

Chapter 16

Proof by Contradiction

16.1 Negation review

Exercise 16.1 Write the negation of the following statements.

(1) $\forall a, b \in \mathbb{R}$, if ab is irrational then a is irrational or b is irrational.

(2) $\forall n \in \mathbb{Z}$, if n^2 is even then n is even.

When proving mathematical statements by contradiction start by assuming that the negation of a conditional statement is true. If you had some trouble with the previous exercise then now is a good time to go back to Chapter 4 to review negation rules. Next, we try to show that the negation is false. The negation of a "For all" statement is a "There exists" statement. To disprove a "There exists" statement you need to show that there won't be any such cases that exist. If the original statement was conditional then the negation will turn it into an "and" statement. This means that any of the parts are fair game to start with. Start with the ones that are easier to define and manipulate. For example, if the statement was "$b^3 + 4$ is even and b is odd" then it would make sense to start with b is odd and use the first part to get a contradiction. A contradiction occurs when you get two things that can't both be true. For example, if you get a number is both odd and even. This type of argument was used by ancient Greeks such as Plato. They called the technique "reductio ad absurdum," which is Latin for "reduction to absurdity."

16.2 Proof by contradiction

Example 16.1 Prove by contradiction: $\forall\, a, b \in \mathbb{R}$, if ab is irrational then a is irrational or b is irrational.

Start by assuming that the negation is true: $\exists\, a, b \in \mathbb{R}$ such that ab is irrational and a is rational and b is rational. We can start with a and b being rational since we are assuming all of these are true. Then $a = \frac{m}{n}$ and $b = \frac{k}{j}$, where m, n, k, j are integers and $n, j \neq 0$. Multiplying we get $ab = \frac{mk}{nj}$. We know mk and nj are integers and that $nj \neq 0$. This means that ab is rational. This is a contradiction because ab can't be both rational and irrational at the same time. This means that the negation has to be false since it led to this contradiction, which means that the original statement is true. Here is the proof written out.

Proof.

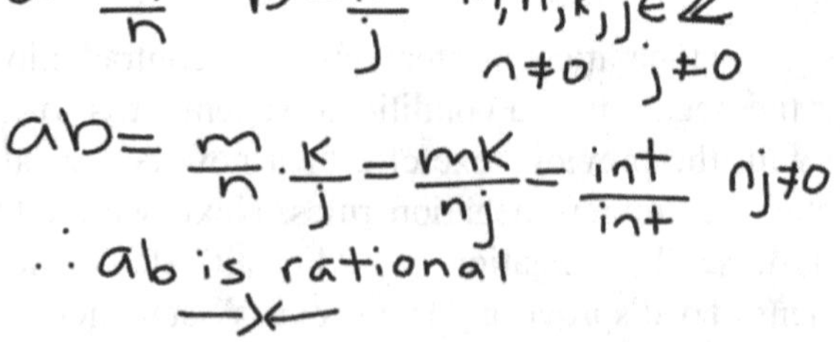

$$a = \frac{m}{n} \quad b = \frac{k}{j} \quad m, n, k, j \in \mathbb{Z}$$
$$n \neq 0 \quad j \neq 0$$
$$ab = \frac{m}{n} \cdot \frac{k}{j} = \frac{mk}{nj} = \frac{int}{int} \quad nj \neq 0$$
$$\therefore ab \text{ is rational}$$
$$\longrightarrow\!\!\times\!\!\longleftarrow$$

Figure 16.1. Example 16.1 proof

The symbol with the two arrows meeting each other stands for a contradiction. You do not have to use this symbol, but it can be a way to show that you got a contradiction. You may have noticed that we proved this same statement in the last chapter by contraposition. In fact, any statement that we can prove by contraposition can also be proved by contradiction.

Exercise 16.2 Look at Example 15.1 and compare it to the example we just did. How are they alike and how are they different?

_____

_____

_____

_____

_____

_____

Example 16.2 Prove by contradiction: $\forall n \in \mathbb{Z}$, if n^2 is even then n is even.

Start by assuming that the negation is true: $\exists n \in \mathbb{Z}$ such that n^2 is even and n is odd. We can start with n being odd since we are assuming both of these are true. Then $n = 2k + 1$, where k is an integer. Square both sides to get

$$n^2 = (2k+1)^2 = 4k^2 + 4k + 1 = 2(2k^2 + 2k) + 1 = 2(\text{integer}) + 1 = \text{odd}. \quad (16.1)$$

This is a contradiction because n^2 can't be both even and odd at the same time. This means that the negation has to be false since it led to this contradiction, which means that the original statement is true. □

Exercise 16.3 The last two examples we have now proved by contraposition and by contradiction. Do you think the two proof methods are interchangeable or can you think of some proofs that will work with one method but not the other?

_____

_____

_____

_____

Exercise 16.4 Explain why the following is not a correct proof.

Prove: $\forall\, n, m \in \mathbb{Z}$, if nm is even then n is even or m is even.

Assume the negation: $\exists\, n, m \in \mathbb{Z}$ such that nm is even and n is odd and m is odd.

Proof. Let $n = 2$ and $m = 5$. Then $nm = 10$ is even. We found an example where nm is even and m is odd but n is not odd so the negation must be false. This means that the original statement is true. $\square$

_____

_____

_____

_____

Exercise 16.5 Now correctly prove $\forall\, n, m \in \mathbb{Z}$, if nm is even then n is even or m is even.

Example 16.3 Prove $\sqrt{2}$ is irrational.

This statement we are not able to prove directly because we do not have a definition for being irrational, other than saying it means not rational. Only knowing that $\sqrt{2}$ cannot be written as a ratio of two integers does not give us a way to represent it. We also can't use proof by contraposition because it is not a conditional statement. Let's try proof by contradiction.

Proof. Start by assuming the negation; assume that $\sqrt{2}$ is rational. Then we can use the definition for rational to write $\sqrt{2} = \frac{a}{b}$ where a and b are integers and $b \neq 0$. We will make one additional assumption that $\frac{a}{b}$ is in lowest terms, i.e. a and b have no common factors. Next, we will square both sides to get

$$2 = \frac{a^2}{b^2} \rightarrow 2b^2 = a^2. \tag{16.2}$$

We know b is an integer, therefore b^2 is also an integer. This means that a^2 is of the form 2(integer), or an even number. Using the result of Example 16.2, we get that a must also be even. By definition $a = 2k$, where k is an integer. We will take this and plug it in for a in the equation we had earlier, $2b^2 = a^2$. This leads to

$$2b^2 = (2k)^2 = 4k^2. \tag{16.3}$$

Dividing both sides by 2 yields

$$b^2 = 2k^2 = 2(\text{integer}). \tag{16.4}$$

This means that b^2 is even. We can then conclude that b is even using Example 16.2 again. So we ended up with a is even and b is even. This is a contradiction because we originally said that a and b have no common factors. If they are both even then they would share a common factor of 2. Our assumption that $\sqrt{2}$ is rational led to this contradiction, which means that it is false and its opposite, $\sqrt{2}$ is irrational, is true. Here is the proof written out.

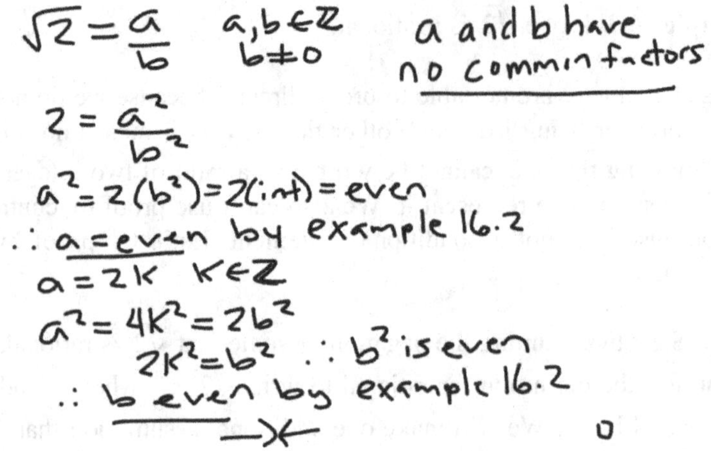

$$\sqrt{2} = \frac{a}{b} \quad a, b \in \mathbb{Z} \quad \underline{a \text{ and } b \text{ have}}$$
$$b \neq 0 \quad \underline{no \text{ common factors}}$$

$$2 = \frac{a^2}{b^2}$$

$$a^2 = 2(b^2) = 2(int) = even$$
$$\therefore \underline{a = even} \text{ by example 16.2}$$
$$a = 2k \quad k \in \mathbb{Z}$$
$$a^2 = 4k^2 = 2b^2$$
$$2k^2 = b^2 \quad \therefore b^2 \text{ is even}$$
$$\therefore \underline{b \text{ even}} \text{ by example 16.2}$$
$$\longrightarrow\!\!\!\!\longleftarrow \qquad \square$$

Figure 16.2.   Proof that square root of 2 is irrational

In this example the contradicting information is underlined. This can be a good technique to help you see where the contradiction occurs.

Exercise 16.6 Complete the following. Use a copy of the proof framework or list your steps following the same format.

(1)   Prove by contradiction: $\forall k \in \mathbb{Z}$, if $3k + 1$ is even then k is odd.
(2)   Prove by contradiction: $\forall a, b \in \mathbb{Z}$, if $a \geq 2$ then a does not divide b or a does not divide $b + 1$.
(3)   Prove by contradiction: $\forall a \in \mathbb{Z}$, if $a^2 - 2a + 7$ is even then a is odd.

Exercise 16.7 Write a reflection on what you learned from any of the mistakes you or a classmate made on any of the proofs you did today.

_____

_____

_____

_____

Exercise 16.8 Fill in the "Proof Methods" table explaining how proof by contradiction works as well as providing an example.

Chapter 16 Homework

Fill in your homework log as you work on the following problems.

(1) Prove or disprove: $\sqrt{6}$ is irrational.

(2) Prove or disprove: $\forall\, a, b \in R$, if a is rational and b is irrational then $a + b$ is irrational.

(3) Prove or disprove: For all integers a and b, $a^2 - 4b \neq 2$.

(4) Prove or disprove: $\forall\, x, y \in Z$, if 3 does not divide xy then 3 does not divide x and 3 does not divide y.

(5) Prove or disprove: $\forall\, a, b, q \in \mathbb{Z}$, if $a = bq + 2$ then $\gcd(a, b) = \gcd(b, 2)$.

Chapter 17

Proof by Induction

17.1 Motivation for induction

Exercise 17.1 Starting with 1, take the sum of any amount of odd numbers going in order. For example, $1 + 3 + 5 = 9$. Do this several times with different amount of odd numbers. What do you notice? Make a conjecture of what the sum of the first n odd positive integers will equal.

Let's look at this problem visually to see what is going on.

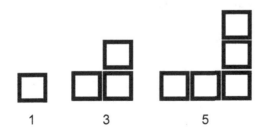

Figure 17.1. Odd numbers

The first three positive odd numbers are shown in a backwards L shape.

Exercise 17.2 Continue this pattern for 7 and then draw and describe what the nth odd number will look like if you follow this pattern.

Next, think of taking the 1 and adding it to the 3 so that it goes in the missing spot to complete the square shape. This shows that when you add the first two positive odd numbers you get 2^2.

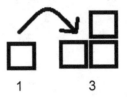

1　　　　　3

Figure 17.2.　Sum of odds becoming square

Exercise 17.3 Take your new 2×2 square and add it to the 5 picture to show that this also forms a square. What have you shown? If you added all of the previous terms to the picture you drew of the nth positive odd number what would the sum become?

This visual argument gives you an intuitive feel for why the sum of the first n odd numbers turns into a square, but to formally prove it we need a new proof technique called mathematical induction. We'll get into the details in a little bit, but first we will look at a few more problems.

Exercise 17.4 Use a similar visual method to make a conjecture for what the sum of the first n even numbers is. Again, we will come back to this problem and formally prove it with induction.

We have already proved the following two statements:

(1) For all integers x, if x^2 is even then x is even.
(2) For all integers x, if x^3 is even then x is even.

Exercise 17.5 What methods did we use to prove these two statements? Explain the steps of the proofs.

_____

_____

_____

_____

_____

_____

_____

Exercise 17.6 Given x is an integer, prove for all integers $n \geq 2$, if x^n is even then x is even.

That last exercise was purposely difficult to try to prove a point. It's okay if you weren't able to solve it. It is possible using the Binomial Theorem, but we'll learn a much nicer method, which you have probably guessed by now will be by induction. Rather than jumping right into induction, we looked at these problems first so that you are hopefully convinced how useful proof by induction is and are very curious to see how it works.

17.2 Induction definition

Exercise 17.7 Think of an infinite set of dominoes set up. How could you prove to somebody that you could knock down all of those dominoes?

Figure 17.3. Dominoes

_____

_____

_____

_____

Induction can be used when you want to prove that a statement is true over the set of natural numbers, or a subset of the natural numbers. This means that the statement uses only positive whole numbers. If it involves real numbers then induction will not work. We'll follow the idea of the infinite set of dominoes. You really only need to show two things: that the first domino can be knocked over and that any one domino can knock over the one right after it. With these two true you can prove that the dominoes can fall for as long as you want. The first one can fall. Then we know that the first can knock the second since any one domino can knock the one right after it. Then we also know the second can knock the third, and so on. Proof by induction will work the same way. You will need to prove that the statement is true for the first value. Generally this is 1, but it could start at a later whole number. Then you need to prove what is called the inductive step: If the statement holds for some arbitrary natural number k then it will also be true for the next whole number, $k+1$. Now we will go back to our first exercise and actually prove that the sum of the first n odd positive integers is the nth square number.

17.3   Proof by induction

Example 17.1   Prove: $\forall n \in \mathbb{N}, 1+3+5+\ldots+(2n-1) = n^2$.

Proof. The first step is to prove this is true for the starting value. In this case the starting value is 1 because we are asked to prove it for all natural numbers and 1 is the first natural number.

Prove for $n = 1$: $1 = 1^2$.
The left side and right side are both equal to 1 so this statement holds true for $n = 1$.

Next, we move to the inductive step. Assume the statement is true for $n = k$: $1+3+5+\ldots+(2k-1) = k^2$.

Notice that this statement is the same as the original with all of the n values changed to k values. Nothing else has been changed. Now we need to prove the statement is true for the next value, $n = k+1$. We will write out what we want to prove with $k+1$ plugged in for n. You will notice one other change. On the left side rather than just listing the last term (the $k+1$ term), we will also list the previous term (the k term). The reason for this will make sense once we go through the proof, but the idea is that we want to use what we know (that the statement holds for $n = k$) to prove what we want (that it also holds for $n = k+1$).

Exercise 17.8   Is the following equality true? Explain why or why not.
$1+3+5+\ldots+(2k-1)+(2k+1) = 1+3+5+\ldots+(2k+1)$.

Here is what we want to prove:

$$n = k+1: 1+3+5+\ldots+(2k-1)+(2k+1) = (k+1)^2. \quad (17.1)$$

This is a statement that we want to prove so we do not know if it is true yet. Therefore, don't start with the two sides being equal; that is what you need to prove. Rather, start with one side and eventually prove that it equals the other side.

$$\text{Left side} = \left[1+3+5+\ldots+(2k-1)\right]+(2k+1)$$
$$= k^2+2k+1 = (k+1)^2 = \text{Right side } \square$$

Figure 17.4. Example 17.1 proof

Exercise 17.9 Explain how we can get each step in the final line of this proof.

_____

_____

_____

_____

Exercise 17.10 Now prove that the sum of the first n even positive numbers is equal to n(n+1), i.e. prove: $\forall n \in \mathbb{N}$, $2+4+6+\ldots+(2n) = n(n+1)$.

Exercise 17.11 Use the results of the sum of the first n even and odd positive numbers to conjecture what the sum of the first n positive numbers is $(1+2+\ldots+n)$ and then prove it by induction. One possible solution can be found in Appendix D, but only look at this after attempting the proof yourself.

Exercise 17.12 Explain what is good about the following proof, what is incorrect, and how it can be fixed to be correct. Include a correct proof in your explanation.

Prove: For all natural numbers n, $\frac{1}{2} + \frac{1}{6} + \cdots + \frac{1}{n(n+1)} = \frac{n}{n+1}$.

Proof.

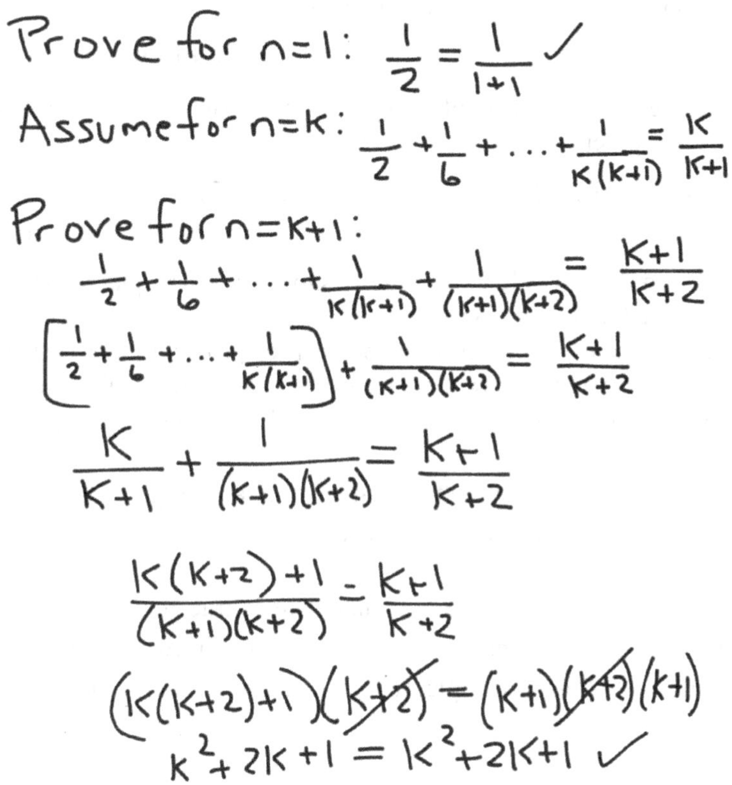

$$\text{Prove for } n=1: \quad \frac{1}{2} = \frac{1}{1+1} \checkmark$$

$$\text{Assume for } n=k: \quad \frac{1}{2} + \frac{1}{6} + \ldots + \frac{1}{k(k+1)} = \frac{k}{k+1}$$

$$\text{Prove for } n=k+1:$$

$$\frac{1}{2} + \frac{1}{6} + \ldots + \frac{1}{k(k+1)} + \frac{1}{(k+1)(k+2)} = \frac{k+1}{k+2}$$

$$\left[\frac{1}{2} + \frac{1}{6} + \ldots + \frac{1}{k(k+1)}\right] + \frac{1}{(k+1)(k+2)} = \frac{k+1}{k+2}$$

$$\frac{k}{k+1} + \frac{1}{(k+1)(k+2)} = \frac{k+1}{k+2}$$

$$\frac{k(k+2)+1}{(k+1)(k+2)} = \frac{k+1}{k+2}$$

$$\left(k(k+2)+1\right)\left(k+2\right) = (k+1)(k+2)(k+1)$$

$$k^2 + 2k + 1 = k^2 + 2k + 1 \checkmark$$

Figure 17.5. Exercise 17.12 proof

We saw in Exercise 17.12 that you are not allowed to use what you want to prove within the proof, but it is okay to put one side of what you want to prove into a more usable form. Don't feel that the proof will always work nicely starting with the left side and ending at the right side. It is perfectly okay to work on both sides as long as you keep those sides separate. We don't know that they are equal until it has been proven. As an example, say you wanted to prove for all natural numbers n,

$$14 + 18 + \ldots + (4n + 10) = 2n(n + 6). \qquad (17.2)$$

In the inductive step we would need to prove that

$$14 + 18 + \ldots + (4(k + 1) + 10) = 2(k + 1)(k + 1 + 6). \qquad (17.3)$$

In this problem it will be helpful to work on the right side and multiply the terms out to get

$$2(k + 1)(k + 1 + 6) = 2(k + 1)(k + 7) = 2(k^2 + 8k + 7) = 2k^2 + 16k + 14.$$
$$(17.4)$$

Then work on the left side to eventually show that it is equal to what the right side ended up being, $2k^2 + 16k + 14$.

This next example comes from our earlier problem. The first three induction proofs we did were to prove equality statements. This next one is to prove a conditional statement. It will work the same way.

Example 17.2 Prove by induction: Given x is an integer, prove for all integers $n \geq 2$, if x^n is even then x is even.

Proof. First we need to prove this statement is true for $n = 2$: if x^2 is even then x is even. We have already proved this in two different ways (see Example 15.2 and Example 16.2) so let's move on to the inductive step.

Assume true for $n = k$: if x^k is even then x is even.
Prove for $n = k + 1$: if x^{k+1} is even then x is even.

Remember that the point of induction is that at some point in proving the statement we will use our assumption. Right now we know that x^{k+1} is even and from that we want to prove that x is even. We also know that $x^{k+1} = (x^k)(x)$. If you look back to Chapter 12 you will see that we have proved the following: $\forall a, b \in \mathbb{Z}$, if ab is even then a or b is even. If you didn't recall proving this previously, you could have proved it now and used it as a lemma to continue with our current proof. By this lemma we now know that x^k is even or x is even. In either case we need to show that x is even. The second case is trivial since that is exactly what it says. The first case comes from our assumption — we said that if x^k is even then x must be even. In either case we have now proven that x must be even. $\square$

Exercise 17.13 Complete the following.

(1) Prove for all integers $n \geq 1$, $8 + 10 + 12 + \cdots + (2n + 6) = n^2 + 7n$.
(2) Prove for all integers $n \geq 1$, $1 + 2 + 2^2 + \cdots + 2^{n-1} = 2^n - 1$.
(3) Prove for all positive integers n, $5^n - 1$ is divisible by 4.
(4) Prove for all integers $n \geq 1$, $3 \mid (n^3 + 2n)$.

Exercise 17.14 Write a reflection on what you learned from any of the mistakes you or a classmate made on any of the proofs you did today.

_____

_____

_____

Exercise 17.15 Fill in the "Proof Methods" table explaining how proof by induction works as well as providing an example.

Chapter 17 Homework

Fill in your homework log as you work on the following problems.

(1) Prove for all positive integers a,b, and n, $(ab)^n = a^n b^n$.

(2) Prove: $\forall a, n, m \in \mathbb{Z}$ with $m \neq 0$, $nm | am$ if and only if $n | a$.

(3) Prove for all integers $n \geq 0$, $8 | (3^{2n} - 1)$.

(4) Prove: $\forall a, b \in \mathbb{Z}$, if $a + b$ is odd then a is odd or b is odd.

(5) Prove for all integers $n \geq 1$, $6 + 10 + 14 + \cdots + (4n + 2) = n(2n + 4)$.

Chapter 18

Proof by Induction Part II

18.1 Helpful properties

In this chapter we will still be using mathematical induction but the statements we will be proving will now be inequalities. The steps are still the same but you will see some differences in how the proof is approached. Before we get started with the proofs we will go over some properties that will be useful while doing these type of proofs.

Property 18.1 $\forall x, y, a \in \mathbb{R}$, if $a > 0$ and $x < y$ then $ax < ay$.

This property states that it is okay to multiply both sides of an inequality by a positive number and the inequality will still hold.

Property 18.2 $\forall x, y, a \in \mathbb{R}$, if $x < y$ then $x + a < y + a$.

This property is similar but for addition.

Property 18.3 $\forall x, y, z \in \mathbb{R}$, if $x < y$ and $y < z$ then $x < z$.

This is the transitive property for inequalities.

Property 18.4 $\forall b, n \in \mathbb{Z}$, $b^{n+1} = b(b^n) = b^n + (b-1)(b^n)$.

The first part of this property uses a rule of exponents that when two terms are multiplying with the same base then you add the exponents. For example, if you have 3^4 you can break it down into

$$3 \cdot 3 \cdot 3 \cdot 3 = 3(3 \cdot 3 \cdot 3) = 3(3^3). \tag{18.1}$$

175

Essentially, one of the 3s got pulled out and the exponent dropped by 1. The second part of this property breaks the expression down further by separating out a copy of the base. Keeping with the same example, $3(3^3)$ means that there are 3 copies of 3^3. Another way to write this is one copy plus two copies, or $3^3 + 2(3^3)$.

Exercise 18.1 Given k is a positive integer, break down 5^{k+1} until you get a 5^k term by itself.

Property 18.5 $\forall n \in \mathbb{Z}$, if $n \geq 0$ then $n! = n(n-1)!$

We can see that $5! = 5 \cdot 4 \cdot 3 \cdot 2 \cdot 1 = 5(4 \cdot 3 \cdot 2 \cdot 1) = 5(4!)$. This property gives a way to pull the highest value out of the factorial.

18.2   Proof by induction with inequalities

Example 18.1 Prove: $\forall n \in \mathbb{N}$, $1 + 3n \leq 4^n$.

Proof. Prove true for $n = 1$: $1 + 3(1) \leq 4^1$.
$1 + 3(1) = 4 \leq 4$ is true.

Assume for $n = k$: $1 + 3k \leq 4^k$.
Prove for $n = k + 1$:

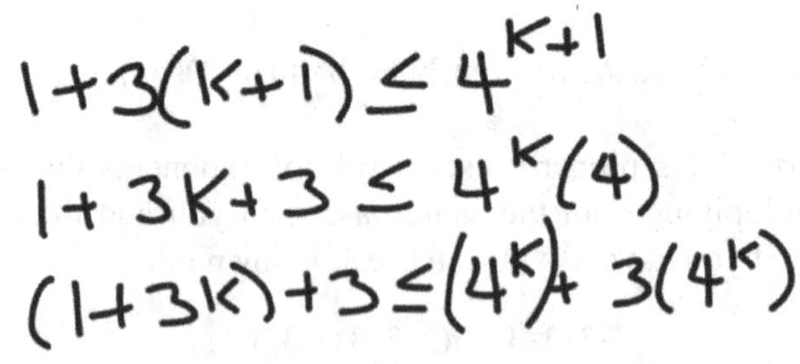

$$1 + 3(k+1) \leq 4^{k+1}$$
$$1 + 3k + 3 \leq 4^k(4)$$
$$(1 + 3k) + 3 \leq (4^k) + 3(4^k)$$

Figure 18.1.   Example 18.1 inductive step

A good suggestion for these type of proofs is to put what you want to prove into a more usable form. That means to make it look similar to your assumption. In this example you can now see that to get from the assumption to the statement that you want to prove, 3 is added on the left and 3 copies of 4^k are added on the right. This is where we eventually want to end up, but for now let's go back to the assumption since we know that is true.

$$1 + 3k \le 4^k. \tag{18.2}$$

From here we will match one side to the statement we want to prove. It doesn't matter which you choose. You could add 3 to both sides of your inequality to match the left side of the prove statement or you could add $3(4^k)$ to match the right side of the prove statement. I chose to add 3.

$$1 + 3k + 3 \le 4^k + 3. \tag{18.3}$$

We know that k is a natural number, so it has to be greater than or equal to 1. This means that 4^k is a positive power of 4, so the smallest this value could be is 4. Then it's safe to say that $4^k \ge 1$. Multiplying both sides by 3 yields $3(4^k) \ge 3$. Adding 4^k to both sides yields

$$4^k + 3(4^k) \ge 4^k + 3. \tag{18.4}$$

Now let's put this inequality together with what we got earlier in 18.3:

$$1 + 3k + 3 \le 4^k + 3 \le 4^k + 3(4^k). \tag{18.5}$$

By the transitive property we can leave off the middle step to get

$$1 + 3k + 3 \le 4^k + 3(4^k) \quad \square \tag{18.6}$$

Example 18.2 Prove: $\forall n \in \mathbb{N}$ with $n \ge 4$, $2^n < n!$

Proof. Prove true for $n = 4$: $2^4 < 4!$
$2^4 = 16$ and $4! = 4(3)(2)(1) = 24$. It is true that $16 < 24$.

Assume for $n = k$: $2^k < k!$
Prove for $n = k + 1$:

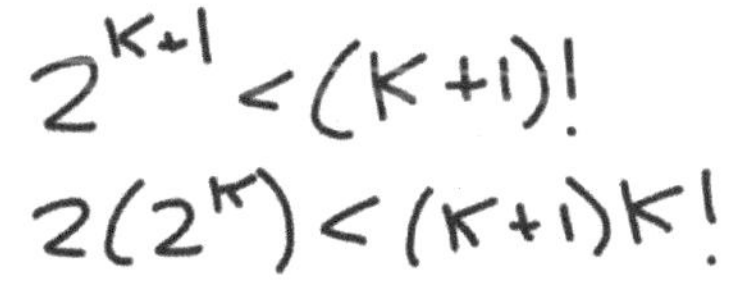

$$2^{k+1} < (k+1)!$$
$$2(2^k) < (k+1)k!$$

Figure 18.2. Example 18.2 inductive step

Just like the last example, we wanted to put the statement to prove into a more useable form. Then start with the assumption.

$$2^k < k! \tag{18.7}$$

You can choose to either multiply both sides by 2 or $(k+1)$. I chose $k+1$.

$$(k+1)2^k < (k+1)k! \tag{18.8}$$

We know that k has to be a natural number ≥ 4 so $k+1 \geq 5$. This means that $k+1 > 2$, which leads to

$$(k+1)2^k > 2(2^k), \text{ or } 2(2^k) < (k+1)2^k. \tag{18.9}$$

Putting this together with our earlier inequality in 18.8 yields

$$2(2^k) < (k+1)2^k < (k+1)k! \tag{18.10}$$

By the transitive property we can leave off the middle step to get

$$2(2^k) < (k+1)k! \tag{18.11}$$

Here is the last step written out.

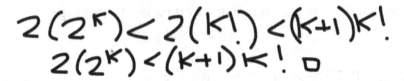

Figure 18.3. Example 18.2 last step in proof

Exercise 18.2 Cindy was asked to prove by induction for all integers $n \geq 2$, $5^n + 9 < 6^n$. She showed the following work. Explain what is good about it, what is incorrect, and how it can be fixed to be correct. Include a correct proof in your explanation.

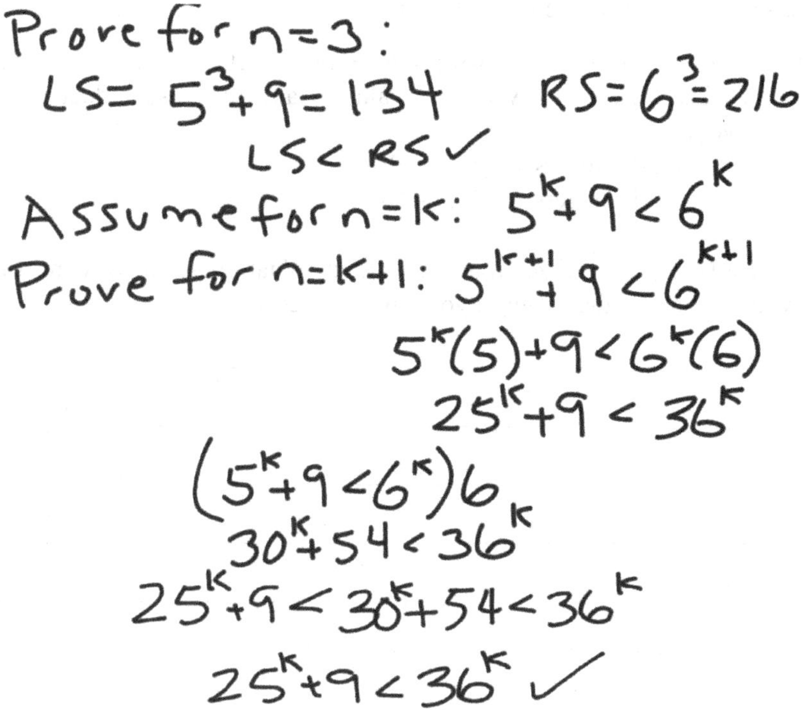

Prove for $n = 3$:

$LS = 5^3 + 9 = 134 \qquad RS = 6^3 = 216$

$\qquad LS < RS \checkmark$

Assume for $n = k$: $\quad 5^k + 9 < 6^k$

Prove for $n = k+1$: $\quad 5^{k+1} + 9 < 6^{k+1}$

$\qquad\qquad 5^k(5) + 9 < 6^k(6)$

$\qquad\qquad 25^k + 9 < 36^k$

$\qquad (5^k + 9 < 6^k)6$

$\qquad 30^k + 54 < 36^k$

$25^k + 9 < 30^k + 54 < 36^k$

$\qquad 25^k + 9 < 36^k \checkmark$

Figure 18.4. Exercise 18.2 proof

Exercise 18.3 Complete the following.

(1) Suppose you want to prove that after a certain value of $n \in \mathbb{N}$, $4n < 2^n$. Figure out what that value is and then use induction to prove it.
(2) Prove for all integers $n > 1$, $3^n > 3n$.
(3) Prove for all integers $n > 1$, $5^n + 9 < 6^n$.

Exercise 18.4 Summarize what we learned. What questions or concerns do you have about the material you have learned so far?

_____

_____

_____

_____

_____

Chapter 18 Homework

Fill in your homework log as you work on the following problems.

(1) Prove for all integers $n \geq 2$, $n^3 > 2n + 1$.

(2) Complete each proof.

   (a) Prove: $\forall a, b, c, m, n \in \mathbb{Z}$, if $a|b$ and $a|c$ then $a|(mb + nc)$.

   (b) Then prove the following corollary: For all integers a,b,c, if $a|b$ and $a|c$ then $a|(b + c)$.

(3) Prove Property 10.2: For all real numbers x, $-|x| \leq x \leq |x|$.

(4) Prove for all integers $n \geq 2$, $n^2 \geq 2n$.

Chapter 19

Calculus Proofs

19.1 Limit proofs

In this chapter we move on to Calculus. There aren't any new proof techniques. Rather, this is a new application to some of the techniques we have learned already. Don't assume any prior limit or derivative rules you may know. Only use rules that are proved in this chapter. The order of problems given is intentional so that you are able to use previously proved problems to help prove later problems. Let f be a function and let a be a real number in the domain of f (i.e. an x-value). The definition for the limit of $f(x)$ as x approaches a equals L is as follows.

Definition 19.1 $\lim_{x \to a} f(x) = L$, means if $\forall \varepsilon > 0$, $\exists \delta > 0$ such that if $0 < |x - a| < \delta$ then $|f(x) - L| < \varepsilon$.

Think of the delta and epsilon values as small, positive values. The idea that this definition conveys is that given a small interval around the y value $f(x)$ you can find a corresponding small interval around the x value. Using the result of Example 12.1 we can re-write $|x - a| < \delta$ to be $-\delta < x - a < \delta$. Next, add a to all terms to get $a - \delta < x < a + \delta$. Notice how this shows an interval around the x value.

$$a - \delta \qquad a \qquad a + \delta$$

Similarly, $|f(x) - L| < \varepsilon$ can be written as an interval around the y value, which is the limit L, $L - \varepsilon < f(x) < L + \varepsilon$.

Think of these proofs as a game; you give me an epsilon and I need to give you a corresponding delta. The way we will figure out this delta is using the Forward-Backward method.

Example 19.1 Let $f(x) = 3x + 5$. Prove: $\lim_{x \to 2} f(x) = 11$.

First, put this in useable form by applying the definition. We want to prove:

$\forall \epsilon > 0$, $\exists \delta > 0$ such that if $0 < |x - 2| < \delta$ then $|(3x + 5) - 11| < \epsilon$.

We will start with scratch work.

$$|f(x) - L| < \epsilon$$
$$|3x + 5 - 11| < \epsilon$$
$$|3x - 6| < \epsilon$$
$$3|x - 2| < \epsilon$$
$$|x - 2| < \epsilon/3$$

Figure 19.1. Example 19.1 scratch work

Based on what we showed here, if we start with $\delta = \frac{\epsilon}{3}$ then the proof will work out.

Proof:

$$|x - a| < \delta$$
$$\text{choose } \delta = \epsilon/3$$
$$|x - 2| < \epsilon/3$$
$$3|x - 2| < \epsilon$$
$$|3x - 6| < \epsilon$$
$$|3x + 5 - 11| < \epsilon$$
$$|f(x) - L| < \epsilon \quad \square$$

Figure 19.2. Example 19.2 proof

That is the end of the proof but let's discuss this example further to better understand what is happening. Based on our work, given any epsilon value we can divide it by 3 to get its corresponding delta value that makes the definition true. If we choose epsilon = 0.3 then the corresponding delta will be 0.1.

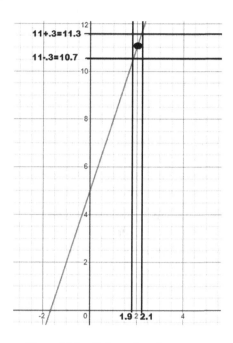

Figure 19.3.   Delta and epsilon ranges

19.2   Exercises

Exercise 19.1 Let $f(x) = \begin{cases} 2x + 3 & \textit{if } x > 5 \\ 7 & x = 5 \\ 3x - 2 & x < 5 \end{cases}$ Prove: $\lim_{x \to 5} f(x) = 13$.

Hint: You need to show the limit from the left and the limit from the right are both 13.

Exercise 19.2 Prove for all real numbers c: $\lim_{x \to a} c = c$.

Exercise 19.3 Prove: if $\lim_{x \to a} f(x) = K$ and $\lim_{x \to a} g(x) = L$ then $\lim_{x \to a} [f(x) + g(x)] = K + L$ (sum rule).

Hint: Use the triangle inequality.

There is a similar rule for the product of limits. This one is a bit harder to prove so we won't go through it, but you can use the result: $\lim_{x \to a} [f(x)g(x)] = KL$.

Definition 19.2 We say f is continuous at a if $\forall \varepsilon > 0$, $\exists \delta > 0$ such that if $|x - a| < \delta$ then $|f(x) - f(a)| < \varepsilon$.

Exercise 19.4

(1) Show the following function is continuous for all real numbers: $f(x) = 3x + 5$.

(2) Show $h(x)$ is continuous at $a = 3$.

$$h(x) = \begin{cases} x & \textit{if } x \geq 3 \\ 2x - 3 & \textit{if } x < 3 \end{cases}$$

Definition 19.3 Alternate definition for continuous: If f is continuous at the point a then $\lim_{x \to a} f(x) = f(a)$.

Exercise 19.5 Explain why these two definitions for continuous at a point a are equivalent.

Definition 19.4 If f is a function and a is a point in the domain of f we say f is differentiable at a if $\lim_{x \to a} \frac{f(x) - f(a)}{x - a}$ exists. In this case we let f'(a) denote the derivative at a, i.e. $f'(a) = \lim_{x \to a} \frac{f(x) - f(a)}{x - a}$.

Exercise 19.6 Let $f(x) = 2x + 3$. Use the derivative definition to find $f'(3)$.

Exercise 19.7 Let f and g be continuous and differentiable functions. Prove the product rule using the definition for derivative: $[f(a)g(a)]' = f'(a)g(a) + g'(a)f(a)$.

Hint: Work with both sides separately and show they equal to the same term.

Exercise 19.8 Use the definition of derivative to find $f'(x)$.

(1) $f(x) = b$ where $b \in \mathbb{R}$.

(2) $f(x) = bx + c$ where $b, c \in \mathbb{R}$.

Exercise 19.9 Use induction to prove the power rule: For all $n \geq 1$, if $f(x) = ax^n$ then $f'(x) = anx^{n-1}$.

Exercise 19.10 Write a reflection on what you learned from any of the mistakes you or a classmate made on any of the proofs you did today.

_____

_____

_____

_____

_____

_____

_____

Chapter 20

Mixed Review

20.1 Review problems

In this chapter you will find a set of problems you can work on for review that come from any proof technique we have learned. Following the questions are a set of possible solutions. The best way to use this chapter for review is to try the problems out yourself first before looking at the solutions. As we discussed in Chapter 6, if you flip right to the solutions and convince yourself that they make sense and you understand them, it will give you an illusion of mastery.

(1) If n is a positive integer, find the possible values of gcd $(n, n + 10)$.

(2) Prove: for all integers a and b, if b does not equal zero then $\gcd(0, b) = |b|$.

(3) Prove: The product of any four consecutive integers is divisible by 4.

(4) Prove: $\forall x \in$ *positive integers*, $x^2 + 1$ has a remainder of 1 or 2 when divided by 4.

(5) Prove: For ints a,b,c, $\gcd(a + cb, b) = \gcd(a, b)$.

(6) Prove the product of two consecutive integers plus the larger of the two integers is a perfect square.

(7) Prove: $10\sqrt{2} + 4$ is irrational.

(8) Prove for all integers a and b, if $a \geq 2$ then a does not divide b or a does not divide $b + 1$.

(9) Prove by induction for all integers $n \geq 7$, $3n^2 > 17n + 10$.

(10) Prove by induction for all integers $n \geq 11$, $n^3 - n > 10n^2$.

   Hint: Use the result of the previous problem.

(11) Prove the quotient rule:

$$\left(\frac{f(x)}{g(x)}\right)' = \frac{g(x)f'(x) - f(x)g'(x)}{(g(x))^2}.$$

(12) If a and b are positive real numbers then prove $a + b \geq 2\sqrt{ab}$.

(13) If a,b, and c are integers such that $a^2 + b^2 = c^2$ then prove at least one of a or b must be even.

(14) Prove for all integers n: if $3|n^2$ then $3|n$.

(15) Prove for all integers x: $3x^2 - 30x + 81 \geq 6$.

(16) Prove: $\forall r \in \mathbb{R}$, r is rational if and only if $r + 2$ is rational.

(17) Prove: If x and y are two integers whose product is even, then at least one of them must be even.

(18) Prove: If x and y are two integers whose product is odd, then both must be odd.

(19) Prove: If a and b a real numbers such that the product ab is an irrational number, then either a or b must be an irrational number.

(20) Prove: If a is an integer, then a is not evenly divisible by 5 if and only if $a^4 - 1$ is evenly divisible by 5.

20.2 Possible solutions

(1) If n is a positive integer, find the possible values of gcd $(n, n + 10)$.

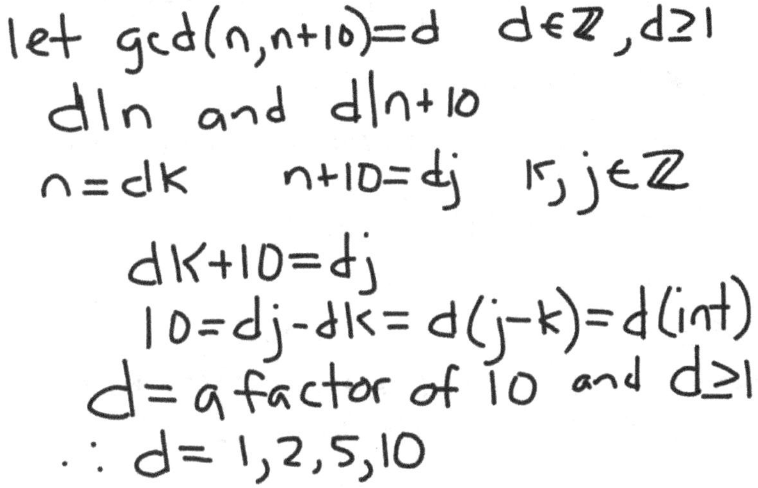

$$\text{let } \gcd(n, n+10) = d \quad d \in \mathbb{Z}, d \geq 1$$
$$d \mid n \text{ and } d \mid n+10$$
$$n = dk \qquad n+10 = dj \qquad k, j \in \mathbb{Z}$$
$$dk + 10 = dj$$
$$10 = dj - dk = d(j-k) = d(\text{int})$$
$$d = a \text{ factor of } 10 \text{ and } d \geq 1$$
$$\therefore d = 1, 2, 5, 10$$

Figure 20.1. 20.1 solution

(2) Prove: for all integers a and b, if b does not equal zero then gcd$(0, b) =$ $|b|$.

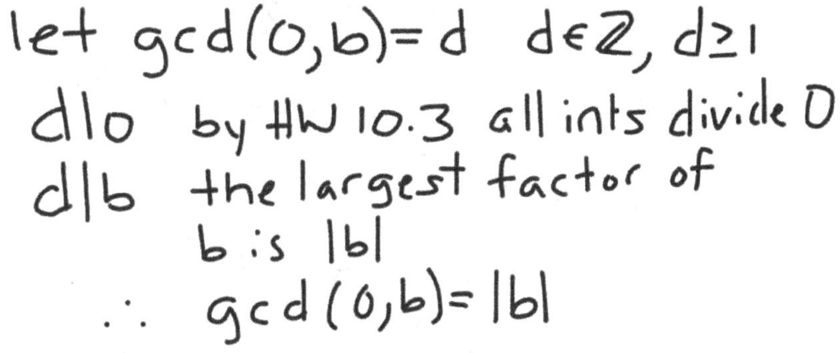

$$\text{let } \gcd(0, b) = d \quad d \in \mathbb{Z}, d \geq 1$$
$$d \mid 0 \quad \text{by HW 10.3 all ints divide 0}$$
$$d \mid b \quad \text{the largest factor of}$$
$$\qquad b \text{ is } |b|$$
$$\therefore \gcd(0, b) = |b|$$

Figure 20.2. 20.2 solution

(3) Prove: The product of any four consecutive integers is divisible by 4.

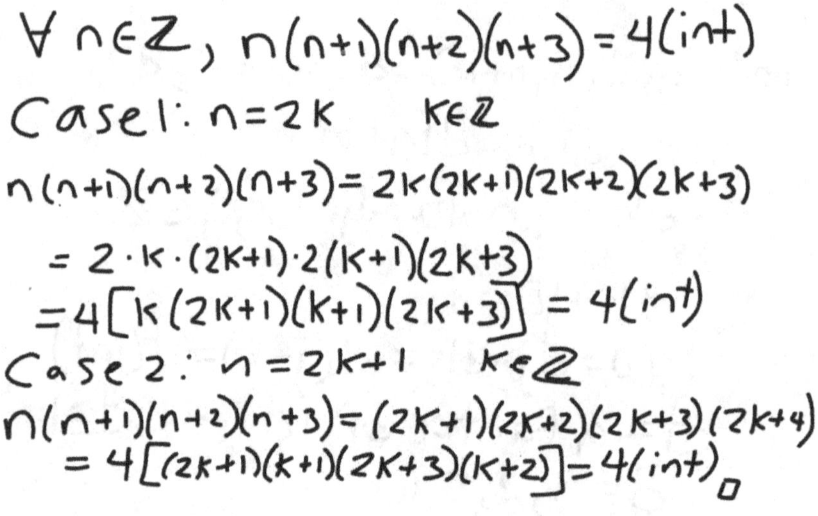

$$\forall n \in \mathbb{Z}, \; n(n+1)(n+2)(n+3) = 4(int)$$

$$\text{Case 1: } n = 2k \qquad k \in \mathbb{Z}$$

$$n(n+1)(n+2)(n+3) = 2k(2k+1)(2k+2)(2k+3)$$

$$= 2 \cdot k \cdot (2k+1) \cdot 2(k+1)(2k+3)$$
$$= 4[k(2k+1)(k+1)(2k+3)] = 4(int)$$

$$\text{Case 2: } n = 2k+1 \qquad k \in \mathbb{Z}$$

$$n(n+1)(n+2)(n+3) = (2k+1)(2k+2)(2k+3)(2k+4)$$
$$= 4[(2k+1)(k+1)(2k+3)(k+2)] = 4(int) \quad \square$$

Figure 20.3. 20.3 solution

(4) Prove: $\forall x \in$ *positive integers*, $x^2 + 1$ has a remainder of 1 or 2 when divided by 4.

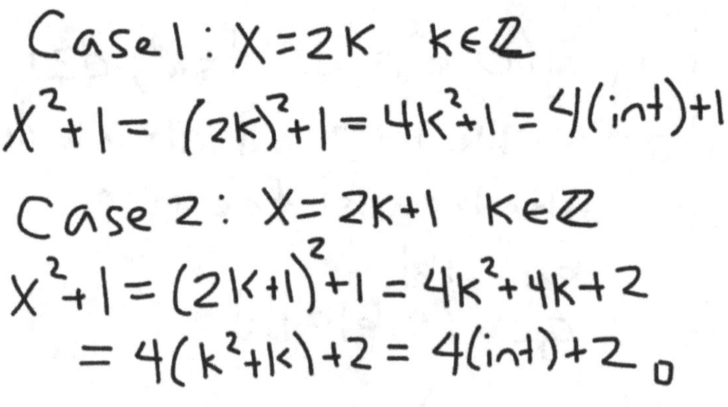

$$\text{Case 1: } X = 2k \qquad k \in \mathbb{Z}$$

$$X^2 + 1 = (2k)^2 + 1 = 4k^2 + 1 = 4(int) + 1$$

$$\text{Case 2: } X = 2k+1 \qquad k \in \mathbb{Z}$$

$$X^2 + 1 = (2k+1)^2 + 1 = 4k^2 + 4k + 2$$
$$= 4(k^2 + k) + 2 = 4(int) + 2 \quad \square$$

Figure 20.4. 20.4 solution

(5) Prove: For ints a,b,c, $\gcd(a + cb, b) = \gcd(a, b)$.

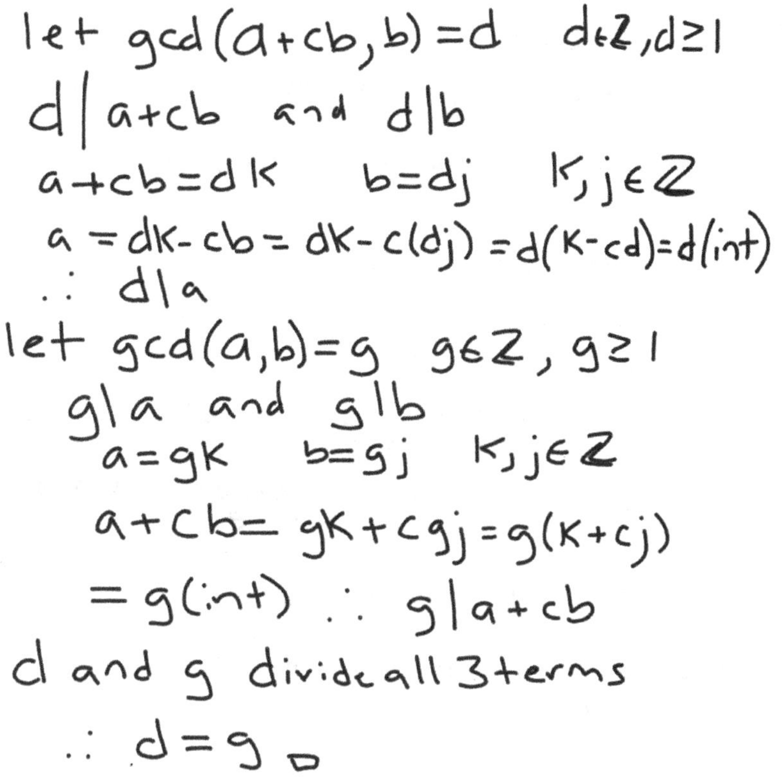

$$\text{let } \gcd(a+cb, b) = d \quad d \in \mathbb{Z}, d \geq 1$$
$$d \mid a+cb \quad \text{and} \quad d \mid b$$
$$a+cb = dk \quad b = dj \quad k, j \in \mathbb{Z}$$
$$a = dk - cb = dk - c(dj) = d(k - cd) = d(int)$$
$$\therefore \; d \mid a$$
$$\text{let } \gcd(a, b) = g \quad g \in \mathbb{Z}, g \geq 1$$
$$g \mid a \quad \text{and} \quad g \mid b$$
$$a = gk \quad b = gj \quad k, j \in \mathbb{Z}$$
$$a + cb = gk + cgj = g(k + cj)$$
$$= g(int) \quad \therefore \; g \mid a + cb$$
$$d \text{ and } g \text{ divide all 3 terms}$$
$$\therefore \; d = g \quad \square$$

Figure 20.5. 20.5 solution

(6) Prove the product of two consecutive integers plus the larger of the two integers is a perfect square.

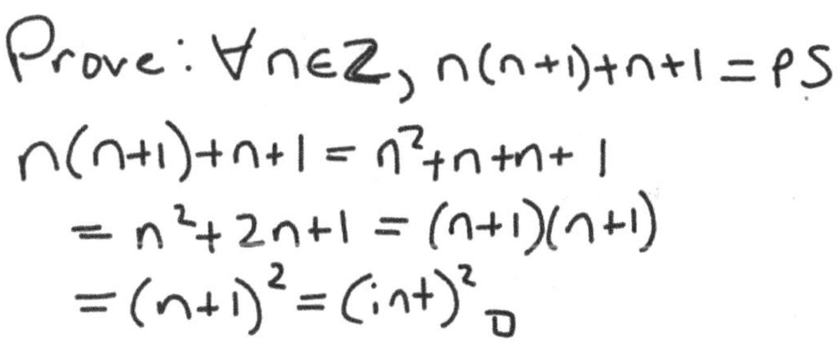

$$\text{Prove}: \forall n \in \mathbb{Z}, \; n(n+1) + n + 1 = PS$$
$$n(n+1) + n + 1 = n^2 + n + n + 1$$
$$= n^2 + 2n + 1 = (n+1)(n+1)$$
$$= (n+1)^2 = (int)^2 \quad \square$$

Figure 20.6. 20.6 solution

(7) Prove: $10\sqrt{2} + 4$ is irrational.

Assume negation:
   $10\sqrt{2} + 4$ is rational

   $10\sqrt{2} + 4 = \dfrac{a}{b}$ $a, b \in \mathbb{Z}$ $b \neq 0$

   $10\sqrt{2} = \dfrac{a}{b} - 4 = \dfrac{a - 4b}{b}$

   $10b\sqrt{2} = a - 4b$
   $\sqrt{2} = \dfrac{a - 4b}{10b} = \dfrac{\text{int}}{\text{int}}$ $10b \neq 0$
   ∴ $\sqrt{2}$ is rational ✗ proved $\sqrt{2}$ is irrational
negation false ∴ original true

Figure 20.7.   20.7 solution

(8) Prove for all integers a and b, if a ≥ 2 then a does not divide b or a does not divide b+1.

CP: $\forall a, b \in \mathbb{Z}$, if $a | b$ and $a | b+1$ then $a < 2$
   $b = ak$
   $b + 1 = aj$ $k, j \in \mathbb{Z}$
   $ak + 1 = aj$
   $1 = aj - ak = a(j-k) = a(\text{int})$
   ∴ $a = \pm 1$ ∴ $a < 2$
CP true ∴ original true

Figure 20.8.   20.8 solution

(9) Prove by induction for all integers $n \geq 7$, $3n^2 > 17n + 10$.

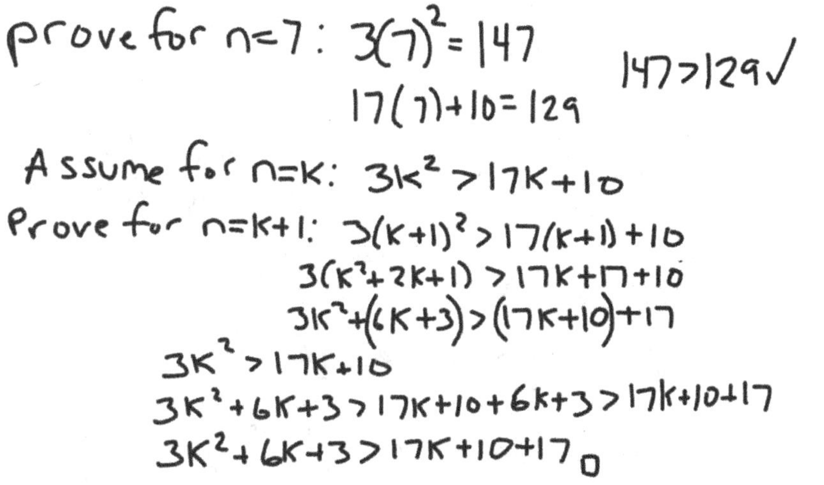

$$\text{Prove for } n=7: \ 3(7)^2 = 147$$
$$17(7)+10 = 129 \qquad 147 > 129 \checkmark$$

$$\text{Assume for } n=k: \ 3k^2 > 17k + 10$$
$$\text{Prove for } n=k+1: \ 3(k+1)^2 > 17(k+1) + 10$$
$$3(k^2 + 2k + 1) > 17k + 17 + 10$$
$$3k^2 + (6k+3) > (17k+10) + 17$$
$$3k^2 > 17k + 10$$
$$3k^2 + 6k + 3 > 17k + 10 + 6k + 3 > 17k + 10 + 17$$
$$3k^2 + 6k + 3 > 17k + 10 + 17 \ \square$$

Figure 20.9. 20.9 solution

(10) Prove by induction for all integers $n \geq 11$, $n^3 - n > 10n^2$.
Hint: Use the result of the previous problem.

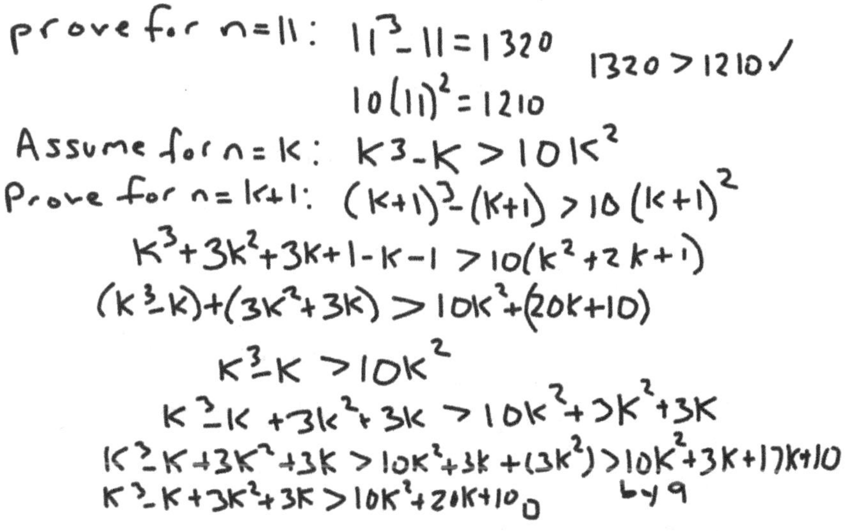

$$\text{Prove for } n=11: \ 11^3 - 11 = 1320$$
$$10(11)^2 = 1210 \qquad 1320 > 1210 \checkmark$$

$$\text{Assume for } n=k: \ k^3 - k > 10k^2$$
$$\text{Prove for } n=k+1: \ (k+1)^3 - (k+1) > 10(k+1)^2$$
$$k^3 + 3k^2 + 3k + 1 - k - 1 > 10(k^2 + 2k + 1)$$
$$(k^3 - k) + (3k^2 + 3k) > 10k^2 + (20k + 10)$$
$$k^3 - k > 10k^2$$
$$k^3 - k + 3k^2 + 3k > 10k^2 + 3k^2 + 3k$$
$$k^3 - k + 3k^2 + 3k > 10k^2 + 3k + (3k^2) > 10k^2 + 3k + 17k + 10$$
$$k^3 - k + 3k^2 + 3k > 10k^2 + 20k + 10 \ \square \qquad \text{by } 9$$

Figure 20.10. 20.10 solution

(11) Prove the quotient rule:

$$\left(\frac{f(x)}{g(x)}\right)' = \frac{g(x)f'(x) - f(x)g'(x)}{(g(x))^2}.$$

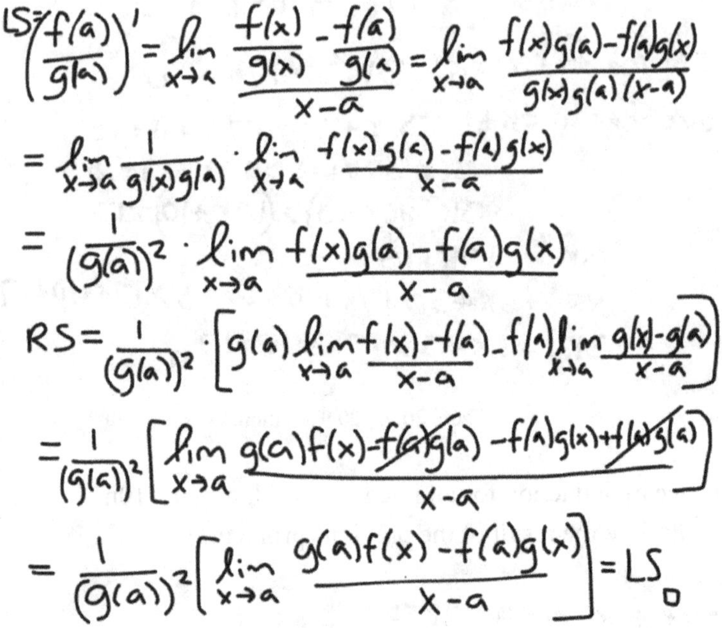

$$LS = \left(\frac{f(a)}{g(a)}\right)' = \lim_{x \to a} \frac{\frac{f(x)}{g(x)} - \frac{f(a)}{g(a)}}{x - a} = \lim_{x \to a} \frac{f(x)g(a) - f(a)g(x)}{g(x)g(a)(x-a)}$$

$$= \lim_{x \to a} \frac{1}{g(x)g(a)} \cdot \lim_{x \to a} \frac{f(x)g(a) - f(a)g(x)}{x - a}$$

$$= \frac{1}{(g(a))^2} \cdot \lim_{x \to a} \frac{f(x)g(a) - f(a)g(x)}{x - a}$$

$$RS = \frac{1}{(g(a))^2} \left[g(a) \lim_{x \to a} \frac{f(x) - f(a)}{x - a} - f(a) \lim_{x \to a} \frac{g(x) - g(a)}{x - a} \right]$$

$$= \frac{1}{(g(a))^2} \left[\lim_{x \to a} \frac{g(a)f(x) - f(a)g(a) - f(a)g(x) + f(a)g(a)}{x - a} \right]$$

$$= \frac{1}{(g(a))^2} \left[\lim_{x \to a} \frac{g(a)f(x) - f(a)g(x)}{x - a} \right] = LS \quad \square$$

Figure 20.11. 20.11 solution

(12) If a and b are positive real numbers then prove $a + b \geq 2\sqrt{ab}$.

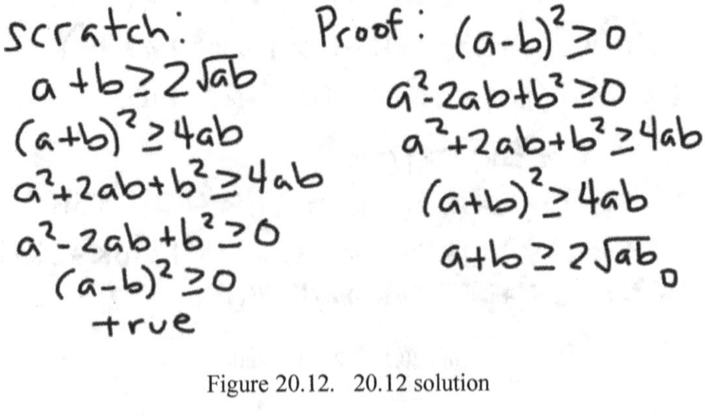

scratch:

$a + b \geq 2\sqrt{ab}$

$(a+b)^2 \geq 4ab$

$a^2 + 2ab + b^2 \geq 4ab$

$a^2 - 2ab + b^2 \geq 0$

$(a-b)^2 \geq 0$

true

Proof: $(a-b)^2 \geq 0$

$a^2 - 2ab + b^2 \geq 0$

$a^2 + 2ab + b^2 \geq 4ab$

$(a+b)^2 \geq 4ab$

$a + b \geq 2\sqrt{ab} \quad \square$

Figure 20.12. 20.12 solution

(13) If a, b, and c are integers such that $a^2 + b^2 = c^2$ then prove at least one of a or b must be even.

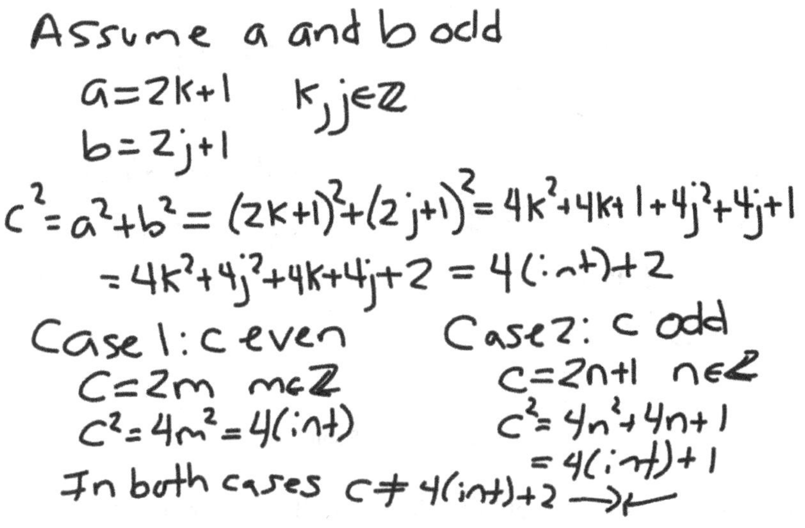
Assume a and b odd
a=2k+1 k,j∈Z
b=2j+1
c²=a²+b²=(2k+1)²+(2j+1)²=4k²+4k+1+4j²+4j+1
=4k²+4j²+4k+4j+2=4(int)+2
Case 1: c even
c=2m m∈Z
c²=4m²=4(int)
Case 2: c odd
c=2n+1 n∈Z
c²=4n²+4n+1
=4(int)+1
In both cases c≠4(int)+2 →←

Figure 20.13. 20.13 solution

(14) Prove for all integers n: if $3|n^2$ then $3|n$.

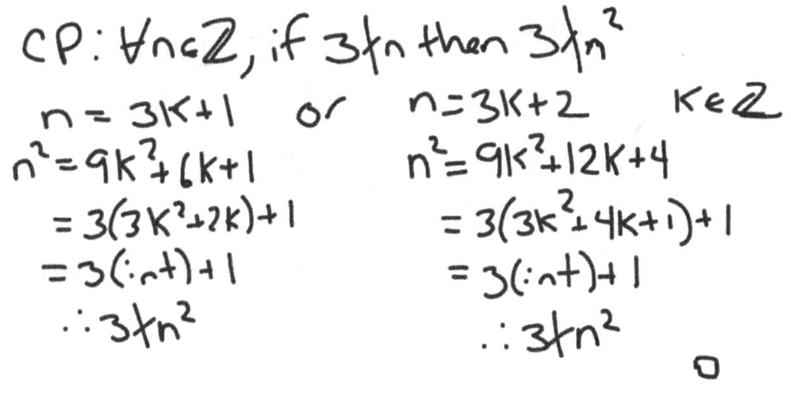
CP: ∀n∈Z, if 3∤n then 3∤n²
n=3k+1 or n=3k+2 k∈Z
n²=9k²+6k+1
=3(3k²+2k)+1
=3(int)+1
∴3∤n²
n²=9k²+12k+4
=3(3k²+4k+1)+1
=3(int)+1
∴3∤n²
□

Figure 20.14. 20.14 solution

(15) Prove for all integers x: $3x^2 - 30x + 81 \geq 6$.

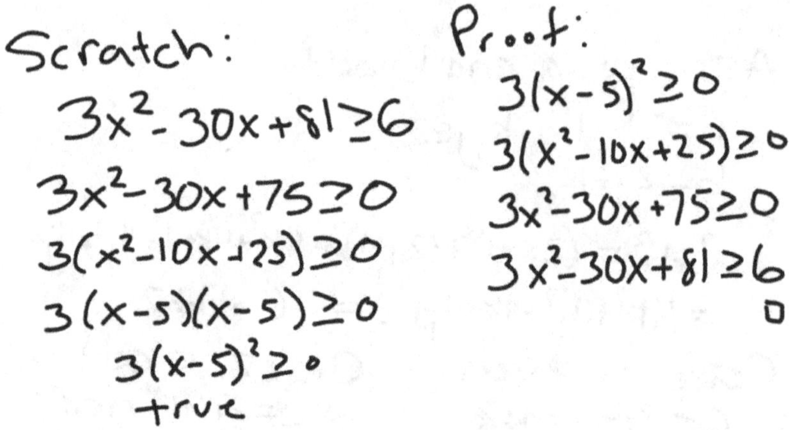

Scratch:

$3x^2 - 30x + 81 \geq 6$

$3x^2 - 30x + 75 \geq 0$

$3(x^2 - 10x + 25) \geq 0$

$3(x-5)(x-5) \geq 0$

$3(x-5)^2 \geq 0$

true

Proof:

$3(x-5)^2 \geq 0$

$3(x^2 - 10x + 25) \geq 0$

$3x^2 - 30x + 75 \geq 0$

$3x^2 - 30x + 81 \geq 6$

□

Figure 20.15. 20.15 solution

(16) Prove: $\forall r \in \mathbb{R}$, r is rational if and only if $r + 2$ is rational.

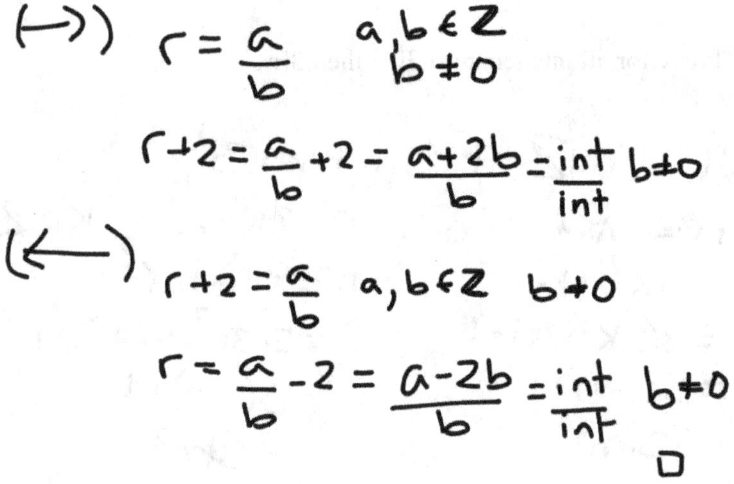

$(\rightarrow))$ $r = \dfrac{a}{b}$ $a, b \in \mathbb{Z}$

$b \neq 0$

$r + 2 = \dfrac{a}{b} + 2 = \dfrac{a + 2b}{b} = \dfrac{int}{int}$ $b \neq 0$

$(\leftarrow)$ $r + 2 = \dfrac{a}{b}$ $a, b \in \mathbb{Z}$ $b \neq 0$

$r = \dfrac{a}{b} - 2 = \dfrac{a - 2b}{b} = \dfrac{int}{int}$ $b \neq 0$

□

Figure 20.16. 20.16 solutions

(17) Prove: If x and y are two integers whose product is even, then at least one of them must be even.

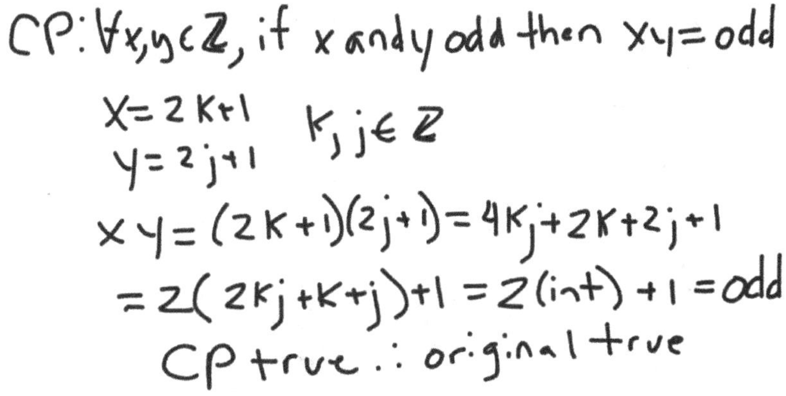

$CP: \forall x, y \in \mathbb{Z},$ if x and y odd then $xy =$ odd

$X = 2K+1 \quad K, j \in \mathbb{Z}$
$Y = 2j+1$

$xy = (2k+1)(2j+1) = 4Kj + 2K + 2j + 1$

$= 2(2Kj + K + j) + 1 = 2(int) + 1 = odd$

CP true $\therefore$ original true

Figure 20.17. 20.17 solution

(18) Prove: If x and y are two integers whose product is odd, then both must be odd.

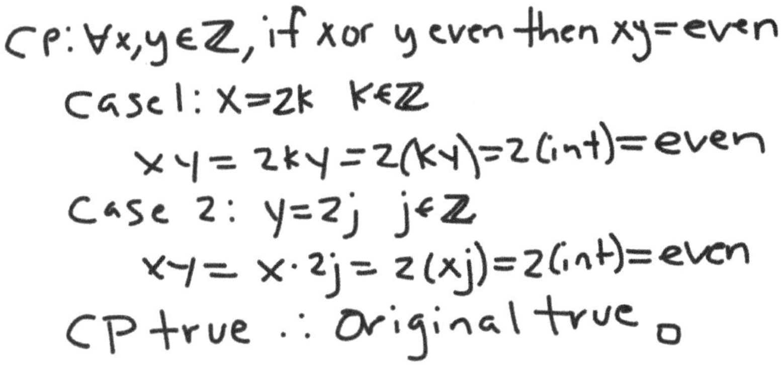

$CP: \forall x, y \in \mathbb{Z},$ if x or y even then $xy =$ even

Case 1: $X = 2k \quad K \in \mathbb{Z}$

$xy = 2ky = 2(ky) = 2(int) = even$

Case 2: $y = 2j \quad j \in \mathbb{Z}$

$xy = x \cdot 2j = 2(xj) = 2(int) = even$

CP true $\therefore$ Original true $\square$

Figure 20.18. 20.18 solution

(19) Prove: If a and b a real numbers such that the product ab is an irrational number, then either a or b must be an irrational number.

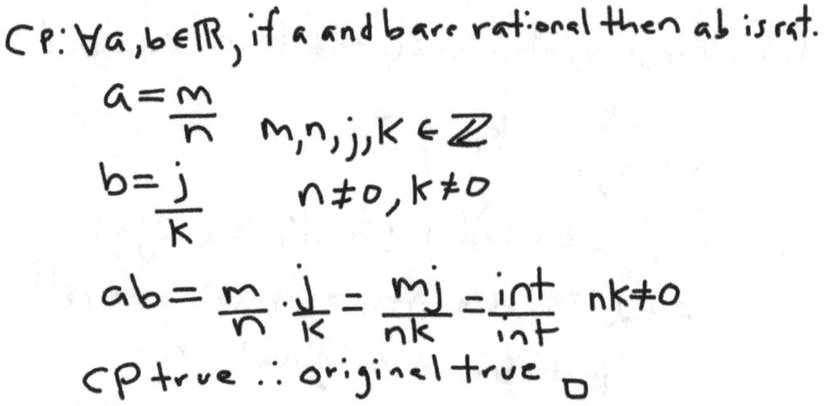

CP: $\forall a, b \in \mathbb{R}$, if a and b are rational then ab is rat.

$$a = \frac{m}{n} \quad m,n,j,k \in \mathbb{Z}$$

$$b = \frac{j}{k} \quad n \neq 0, k \neq 0$$

$$ab = \frac{m}{n} \cdot \frac{j}{k} = \frac{mj}{nk} = \frac{int}{int} \quad nk \neq 0$$

CP true ∴ original true □

Figure 20.19.　20.19 solution

(20) Prove: If a is an integer, then a is not evenly divisible by 5 if and only if $a^4 - 1$ is evenly divisible by 5.

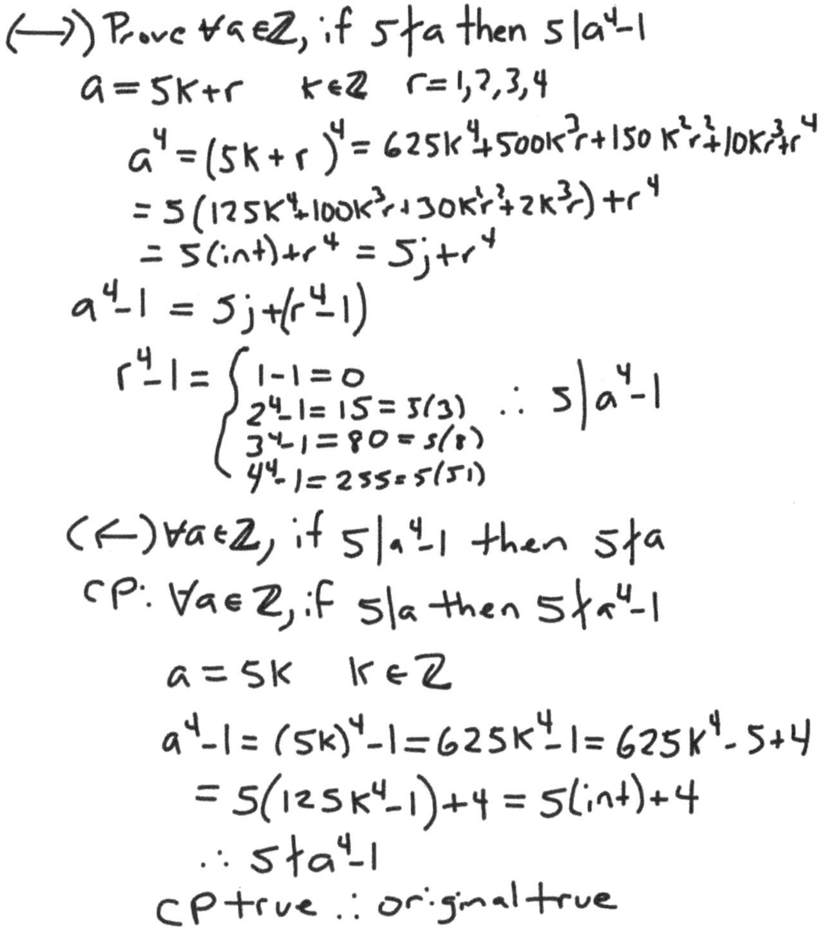

$(\longrightarrow)$ Prove $\forall a \in \mathbb{Z}$, if $5 \nmid a$ then $5 \mid a^4 - 1$

$a = 5k + r \qquad k \in \mathbb{Z} \qquad r = 1, 2, 3, 4$

$a^4 = (5k + r)^4 = 625k^4 + 500k^3 r + 150 k^2 r^2 + 10kr^3 + r^4$

$\quad = 5(125k^4 + 100k^3 r + 30k^2 r^2 + 2kr^3) + r^4$

$\quad = 5(\text{int}) + r^4 = 5j + r^4$

$a^4 - 1 = 5j + (r^4 - 1)$

$r^4 - 1 = \begin{cases} 1 - 1 = 0 \\ 2^4 - 1 = 15 = 5(3) \\ 3^4 - 1 = 80 = 5(16) \\ 4^4 - 1 = 255 = 5(51) \end{cases} \quad \therefore \quad 5 \mid a^4 - 1$

$(\longleftarrow) \forall a \in \mathbb{Z}$, if $5 \mid a^4 - 1$ then $5 \nmid a$

CP: $\forall a \in \mathbb{Z}$, if $5 \mid a$ then $5 \nmid a^4 - 1$

$a = 5k \qquad k \in \mathbb{Z}$

$a^4 - 1 = (5k)^4 - 1 = 625k^4 - 1 = 625k^4 - 5 + 4$

$\quad = 5(125k^4 - 1) + 4 = 5(\text{int}) + 4$

$\therefore 5 \nmid a^4 - 1$

CP true $\therefore$ original true

Figure 20.20. 20.20 solution

100# Task Activity Sheet

37 77 1   81   **34** $\frac{90}{3}$ 2•3

    60+5 18-1   70+4   42

  13   7•3   54

73   33 29   46   42+20 82

  25 41 72+17 25+25 86   25-3 66 10

87-2 93 20+25 5 69 90 2 **26** 18

4+5   36+13 57   12+2 70 94 38

56-3   61 91+6 **60+18** 50+7+1

  4•4 20+20 92 48   43 98 75 50+5

    20   40•2 80+4   47   3+4

32 54-2 8 60   3 10+57   91

100   72 30-6 76 39 $\frac{30}{2}$ 95 7•5 31

2+2 96 6•6 68   60-1   63   83

  20+8   100-1 51   23

64 88 $\frac{24}{2}$ 56 44 11   19 68+3

        79 87

205

Appendix B

Answers for Hiking Activity

According to the United Hiking Society, the ranking of importance for these items are given in the table below.

Table B.1. Answers for hiking activity

Item	Expert	Reason
Food	3	Humans can go for more than 3 weeks without food, but food will be important since you are doing strenuous exercise and for comfort reasons.
Headlamp with batteries	5	Essential for seeing at night. A headlamp is preferred over a flashlight because it leaves your hands free.
First aid kit	7	In case you get injured.
Extra clothes	9	Rain gear and layers can help you adjust to different weather.
Lighter	6	To make a fire to protect against cold, signal for help, or to cook food.
Water	2	Humans can go 100 hours without water but since you will be hiking you will need to stay hydrated.
Compass	1	Extremely important if you get lost. It does not rely on batteries and does not weigh much. In the wilderness you will not be able to view the sky for other clues of your location.
Shelter	8	Will protect against wind and rain if you get stranded or injured, but not too helpful since your plan is to go hiking.
Sunscreen	10	Since you are hiking in the wilderness you will be covered by trees and will not be exposed to the sun.
Knife	4	Good for food preparation, repairing your gear, and first aid.

Escape Room

Save Luna

For this escape room split students into 4 groups. To start off, give each group one copy of the first puzzle.

Your precious dog Luna has been dognapped. The thieves have set up a series of riddles you must solve in order to see her again. Here is the first one.

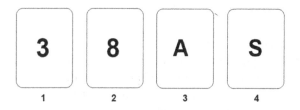

Figure C.1. Cards for escape room

Each card has a number on one side, and a letter on the other. Which card(s) must be turned over to test the idea that if a card shows an even number on one side, then its opposite side is a vowel? Use a 0 if the spot does not need to be turned over and 1 if it does. For example, if you think cards 1, 2, and 3 should be turned over then your code is 1 1 1 0. You only get one shot at unlocking the lock; otherwise, it will self-destruct.

Work in your groups to come up with a solution then send a representative to the thinking desk. Each representative should share their group's answer and try to reach a consensus. If there is a disagreement the representative may go back to their group to discuss. Once all groups agree on an answer then try the code on the lock. The timer has been set to 5 minutes and a final decision must be made by the time it counts down.

You will need to have a lock that has 4 digit spots. Set it to unlock at 0101. Put this lock on a small box which contains the next set of clues. For this next puzzle students will solve a riddle that was posed in *Harry Potter and the Sorcerer's Stone* [1998]. From the book information it is not possible to find a solution because the reader needs to have a visual of what the bottles look like and one is not provided. To make this riddle solvable I 3D printed a set of bottles (see https://www.thingiverse.com/thing:1708846). In the box include one set of bottles for each group as well as a copy of the riddle for each group. I also printed a larger set of the bottles and hid the next clue in the correct bottle.

Danger lies before you, while safety lies behind,
One of us will help you, whichever you would find,
One among us seven will let you move ahead,
Another will transport the drinker back instead,
Two among our number hold only nettle wine,
Three of us are killers, waiting hidden in line.
Choose, unless you wish to stay here forevermore,
To help you in your choice, we give you these clues four:

First, however slyly the poison tries to hide
You will always find some on nettle wine's left side;
Second, different are those who stand at either end,
But if you would move onward, neither is your friend;
Third, as you see clearly, all are different size,
Neither dwarf nor giant holds death in their insides;
Fourth, the second left and second on the right
Are twins once you taste them, though different at first sight. [Rowling, 1998, p. 285].

Which bottle moves you ahead?
Use these practice bottles with your group and send a different representative to the thinking desk. Use the same procedure as riddle 1. Once you have a solution lift the correct bottle in the larger set to reveal the next clue.

Inside the bottle I put 4 small slips of paper. Each has one line of the following riddle.

Vr forvh duh brx
Wr ilqglqj oxqd wkh dgruhg
Doo brx kdyh wr gr
Lv orrn xqghu wkh nhberdug

I also give out decoder rings (see https://www.thingiverse.com/thing: 1706210), though these are not necessary. The clues are written using a Caesar Cipher with a shift of 3 and translate to:

So close are you
To finding Luna the adored
All you have to do
Is look under the keyboard

Taped in an envelope under the keyboard in the classroom is the key to a diary. Written on the first page of the diary is the following:

Hurry, Luna is getting hungry! Each group should pick one riddle to solve. The code for the small lockbox is your solutions (in order).

On a separate page each the following four riddles are written.

(1) If two people can paint two rooms in two days then how many days does it take one person to paint one room?

(2) The students in a particular class are playing a game. The students are split in two teams: Team Red and Team Black. Students in Team Red always tell the truth, students in Team Black always lie. Anna, Daniel, and Claire are three students in the class. The teacher asks Anna: "What team are you?" Anna mumbles something, so the teacher then asks Daniel: "What did Anna say?" Daniel says: "Anna said she is Team Red." Claire then says: "Anna said she is Team Black." What team is Daniel in?

(3) A farmer accidently left a bucket of apples out. A cow came by and ate 1/6 of the apples. Then a pig came by and ate 1/5 of the remaining apples. Next, a goat came by and ate 1/4 of the remaining apples. Next, a horse ate 1/3 of the remaining apples. Finally, a sheep ate 1/2 of the remaining apples. When the farmer came back

to get the bucket there were 3 apples left. How many apples were originally in the bowl? Use the ones digit only.

(4) Albert and Bernard just befriended Cheryl and they want to know when her birthday is. She gives them a list of 10 possible days: May 15, May 16, May 19, June 17, June 18, July 14, July 16, August 14, August 15, and August 17. She tells Albert the month she is born and Bernard the day she is born.

Albert: I don't know when Cheryl's birthday is but neither does Bernard.

Bernard: At first I didn't know when Cheryl's birthday was but I do now.

Albert: Then I also know when it is.

When is Cheryl's birthday? Use the first letter of the month and the last digit of the day.

For this lock you will need 5 wheels. Three will have digits, one with letters, and one with colors. Set the solution to 2(color red)8J6. The lock will be on another box. Inside the box is a stuffed animal dog along with the message: Congratulations, you saved Luna with your impressive problem solving skills!

Materials needed:

- Four copies of the first riddle
- Lock with code 0101 on small box with 4 sets of Harry Potter bottles in them and 4 copies of the riddle
- Big set of Harry Potter bottles with the 4 secret messages hidden inside the "move ahead" bottle
- 4 decoder rings (optional)
- Diary with lock and key with riddles written in them
- Lock with code 2(color red)8J6 on large box with stuffed animal dog and congrats note

I bought a starter kit from Breakout EDU but it is also possible to get together your own materials.

Proof for Exercise 17.11

We have to think of two cases: if n is even and if n is odd. When n is even you will add the same amount of terms from each list. For example, if $n = 2$ (i.e. you want to add the first two natural numbers) you sum the number 1 from the odd list and the number 2 from the even list to get $1 + 2 = 3$. If $n = 4$ you sum 1 and 3 from the odd list and 2 and 4 from the even list to get $1 + 2 + 3 + 4 = 10$. When n is odd you will have one extra term from the odd list. The general formulas are shown in the table.

Table D.1. Number of even and odd terms

n	1	2	3	4	5	6	n (if n is even)	n (if n is odd)
# terms from odd list	1	1	2	2	3	3	$\dfrac{n}{2}$	$\dfrac{n+1}{2}$
# terms from even list	0	1	1	2	2	3	$\dfrac{n}{2}$	$\dfrac{n+1}{2} - 1$

Let's start with the case where n is even. There are $\frac{n}{2}$ terms from the odd list. Plugging this into the formula we proved for summing odds (Example 17.1) yields $\left(\frac{n}{2}\right)^2$. There are also $\frac{n}{2}$ terms from the even list. Plugging this into the formula we proved for summing evens (Exercise 17.10) yields $\frac{n}{2}\left(\frac{n}{2}+1\right)$. Adding these together yields

$$\left(\frac{n}{2}\right)^2 + \frac{n}{2}\left(\frac{n}{2}+1\right) = \frac{n^2 + n^2}{4} + \frac{n}{2} = \frac{2n^2 + 2n}{4} = \frac{n^2 + n}{2}. \qquad (D.1)$$

For the odd case there are $\frac{n+1}{2}$ terms from the odd list. Plugging this into the formula we proved for summing odds yields $\left(\frac{n+1}{2}\right)^2$. There are $\frac{n+1}{2}-1$ terms from the even list. Plugging this into the formula we proved for summing evens yields $\left(\frac{n+1}{2}-1\right)\left(\frac{n+1}{2}-1+1\right)$. Adding these together yields

$$\left(\frac{n+1}{2}\right)^2+\left(\frac{n+1}{2}-1\right)\left(\frac{n+1}{2}-1+1\right)=\frac{n^2+2n+1}{4}+\left(\frac{n+1}{2}-1\right)\left(\frac{n+1}{2}\right)=$$

$$\frac{n^2+2n+1}{4}+\left(\frac{n^2+2n+1}{4}-\frac{n+1}{2}\right)=\frac{2n^2+4n+2}{4}-\frac{2n+2}{4}=\frac{2n^2+2n}{4}=\frac{n^2+n}{2}.$$

$$(D.2)$$

Even though we had to split the problem up into two cases, we ended up with the same result in both of them. We can say that the sum of the first n natural numbers is equal to $\frac{n^2+n}{2}$.

To prove this result by induction, start by proving true for $n=1$.
The sum of the first term is 1. Plugging $n=1$ into the right side yields $\frac{1^2+1}{2}=\frac{2}{1}=1$. This proves the statement is true for $n=1$. Let's move on to the inductive step.

Assume true for $n=k$: $1+2+\dots+k=\frac{k^2+k}{2}$.
Prove true for $n=k+1$: $1+2+\dots+k+(k+1)=\frac{(k+1)^2+k+1}{2}$.

Left side $=1+2+\dots+k+(k+1)=[1+2+\dots+k]+(k+1)=$
$\frac{k^2+k}{2}+(k+1)=\frac{k^2+k}{2}+\frac{2k+2}{2}=\frac{k^2+3k+2}{2}$.

Right side $=\frac{(k+1)^2+k+1}{2}=\frac{k^2+2k+1+k+1}{2}=\frac{k^2+3k+2}{2}=$ left side . $\square$

Note: In Chapter 5 we discussed a much quicker way that Euler used to get the formula for adding the first n natural numbers. I decided to also show this method since we figured out the sum for even and odd numbers in previous problems.

Selected Proofs from all Chapters

This proof list is included here because it may be useful to use these results in a later proof.

Example 8.1 Any even number times any odd number is even.

Exercise 8.12.1 Any even number plus any odd number is odd.

Exercise 8.12.3 Every integer divides itself.

Homework 8.1 Any even number times any even number is even.

Homework 8.3 For all integers a,b,x, and y, if $a|b$ then $a|(ax+by)$.

Example 9.1 The sum of any two rational numbers is rational.

Exercise 9.7.1 The product of any two rational numbers is rational.

Exercise 9.7.2 The square of any rational number is rational.

Homework 9.1 The sum of any two odd integers is even.

Homework 9.4 $\forall a,b,c \in \mathbb{Z}$, if $a|b$ and $b|c$ then $a|c$.

Exercise 10.9.4 $\forall x,y \in \mathbb{R}$, $|x-y| \leq |x|-|y|$.

Homework 10.2 The product of any rational number and any integer is rational.

Homework 10.3 Every integer divides 0.

Homework 11.4 For all non-zero integers a and b, if x and y are any integers then $\gcd(a,b)\,|\,(ax+by)$.

Example 12.1 $\forall x, a \in \mathbb{R}$, $|x| < a$ *if and only if* $-a < x < a$.

Example 12.2 Triangle Inequality: $\forall x, y \in \mathbb{R}$, $|x+y| \le |x| + |y|$.

Exercise 12.10.1 $\forall a, b \in \mathbb{Z}$, if ab is even then a or b is even.

Exercise 12.10.2 For all real numbers x and y, $|xy| = |x| \cdot |y|$.

Exercise 12.10.3 The product of any two consecutive integers is even.

Homework 12.3 $\forall x, y \in \mathbb{R}$, $|x-y| = |y-x|$.

Homework 12.6 $\forall x, y \in \mathbb{Z}$, if $x^2 + y^2$ is even then $x+y$ is even.

Example 15.1 Prove: $\forall a, b \in \mathbb{R}$, if ab is irrational then a is irrational or b is irrational.

Example 15.2 $\forall n \in \mathbb{Z}$, n is even if and only if n^2 is even.

Homework 15.1 For all integers x, if x^3 is even then x is even.

Exercise 17.11 $1 + 2 + \ldots + n = \frac{n(n+1)}{2}$.

Homework 17.4 $\forall a, b \in \mathbb{Z}$, if $a+b$ is odd then a is odd or b is odd.

Homework 18.2 $\forall a, b, c, m, n \in \mathbb{Z}$, if $a\,|\,b$ and $a\,|\,c$ then $a\,|\,(mb+nc)$.

Homework 18.3 For all real numbers x, $-|x| \le x \le |x|$.

Appendix F

Proof Methods

Proof Method	Explanation	Example
Direct		
Proof by Cases		
Forward-Backward		

Contraposition		
Contradiction		
Induction		

Proof Template

Statement to prove:

If needed re-write in $\forall$, if _____ then _____

What do you know? (if needed put in useable form)

What do you want to prove? (if needed put in useable form)

What method(s) might work?

Doodle (play around with your givens and try to get what you want to prove)

Final Proof:

Homework Log

Date	Time Spent on HW (in minutes)

Date	Time Spent on HW (in minutes)

Bibliography

Ashman, A., and Gillies, R. (1997). Children's cooperative behavior and interactions in trained and untrained work groups in regular classrooms. *Journal of School Psychology*, 35(3), pp. 261–279.

Boaler, J. (2019). *Limitless Mind: Learn, Lead, and Live Without Barriers*. New York, NY: HarperOne.

Brown, P. C., Roediger, H. L. and McDaniel, M. A. (2014). *Make it Stick*. Boston, MA: Belknap Press.

Burns, M. (2007). *About Teaching Mathematics: A k-8 Resource*, 3rd ed. Sausalito, CA: Math Solutions Publications.

Carey, B. (2015). *How we Learn: The Surprising Truth About When, Where, and Why it Happens*. New York, NY: Random House.

Callender, A. A. and McDaniel, M. A. (2009). The limited benefits of rereading educational texts. *Contemporary Educational Psychology*, 34(1), pp. 30–41. http://dx.doi.org/10.1016/j.cedpsych.2008.07.001

Chamberlin, K., Yasué, M. and Chiang, I.-C. A. (2018). The impact of grades on student Motivaton. *Active Learning in Higher Education*. https://doi.org/10.1177/1469787418819728

Deutsch, M. (1949). A theory of cooperation and competition. *Human Relations*, 2, pp. 129–152.

Devlin, K. (2011). *Mathematics Education for a New Era: Video Games as a Medium for Learning*. Boca Raton, FL: CRC Press.

Dewey, J. (1896/1980). *The School and Society*. Chicago, IL: The University of Chicago Press.

Dunlosky, J., Rawson, K. A., Marsh, E. J., Nathan, M. J. and Willingham, D. T. (2013). Improving students' learning with effective learning techniques: Promising directions from cognitive and educational psychology. *Psychological Science in the Public Interest*, 14(1), pp. 4–58. https://doi.org/10.1177/1529100612453266

Dweck, C. (2007). *Mindset: The New Psychology of Success*. New York, NY: Ballantine Books.

Hall, K. G., Domingues, D. A. and Cavazos, R. (1994). Contextual inference effects with skilled baseball players. *Perceptual and Motor Skills*, 78, 835–841.

Johnson, R. T. and Johnson, D. W. (1994). *An Overview of Cooperative Learning*, eds. In Thousand, J. Villa, A. and Nevin, A., "Creativity and collaborative learning," (Brookes Press, Baltimore) pp. 31–44.

Kagan, S. (2001). Structures: Research and rationale. Kagan Online Magazine, Spring 2001. Retrieved from: https://www.kaganonline.com/free_articles/research_and_rationale/282/Kagan-Structures-Research-and-Rationale

Lewin, K. (1935). *A Dynamic Theory of Personality*. New York: McGraw-Hill.

Liljedahl, P. (2016). *Building Thinking Classrooms: Conditions for Problem Solving*, eds. Felmer, P., Kilpatrick, J. and Pekhonen, E., "Posing and Solving Mathematical Problems: Advances and New Perspectives," (Springer, New York) pp. 361–386.

Lockhart, Paul (2009). *A mathematician's Lament: New School Cheats Us Out of Our Most Fascinating and Imaginative Art Form*. New York, NY: Bellevue Literary Press.

NACE. (2017, February 22). *Employers rate competencies, students' career readiness*. Retrieved from https://www.naceweb.org/talent-acquisition/internships/employers-rate-competencies-students-career-readiness/

Oakley, B. (2014). *A Mind for Numbers*. New York, NY: Tarcher/Penguin.

Piaget, J. (1932/1965). *The Moral Judgement of the Child*. New York, NY: Routledge.

Polya, G. (1957). *How to Solve It: A New Aspect of Mathematical Method*, 2nd ed. Princeton, NJ: Princeton University Press.

Robinson, K. and Aronica, L. (2016). *Creative Schools: The Grassroots Revolution That's Transforming Education*. New York, NY: Penguin Books.

Rowling, J. K. (1998). *Harry Potter and the Sorcerer's Stone*. New York, NY: Scholastic Press.

Sanderson, G. (2017, May 4). Limits, L'Hopital's rule, and epsilon delta definitions | Essence of calculus, chapter 7. Retrieved January 9, 2020, from https://www.youtube.com/watch?v=kfF40MiS7zAandfeature=youtu.beandt=4m59s.

Schwartz, L. (2001). *A mathematician grappling with his century*. Basel, Switzerland: Birkhauser.

Slavin, R. E. (1991). Synthesis of research on cooperative learning. *Educational Leadership*, pp. 48(5), 71–82.

Slavin, R. E. (1995). *Cooperative Learning: Theory, Research, and Practice*. Boston: Allyn and Bacon.

Smith, M. K. (2001). Kurt Lewin, groups, experiential learning and action research. *The Encyclopedia of Informal Education*. Retrieved from http://www.infed.org/thinkers/et-lewin.htm

Stahl, R. J. (1994). The essential elements of cooperative learning in the classroom. (Available from the ERIC Document Reproduction Service No. ED 370881).

VanDerWerf, S. (2015, December 7). 100 numbers to get student talking. *Sara VanDerWerf*. https://www.saravanderwerf.com/100-numbers-to-get-students-talking/

Vygotsky, L. S. (1978). *Mind in Society*. Cambridge, MA: Harvard University Press.

Index

Printed in the United States
by Baker & Taylor Publisher Services